T0318759

MODERN PRACTICAL HEALTHCARE ISSUES IN BIOMEDICAL INSTRUMENTATION

MODERN PRACTICAL HEALTHCARE ISSUES IN BIOMEDICAL INSTRUMENTATION

Edited by

DILBER UZUN OZSAHIN

DESAM Research Institute, Near East University, Nicosia, Turkish Republic of Northern Cyprus, Turkey; Department of Biomedical Engineering, Near East University, Nicosia, Turkish Republic of Northern Cyprus, Turkey; Medical Diagnostic Imaging Department, College of Health Sciences, University of Sharjah, Sharjah, United Arab Emirates

ILKER OZSAHIN

DESAM Research Institute, Near East University, Nicosia, Turkish Republic of Northern Cyprus, Turkey; Department of Biomedical Engineering, Near East University, Nicosia, Turkish Republic of Northern Cyprus, Turkey; Brain Health Imaging Institute, Department of Radiology, Weill Cornell Medicine, New York, NY, United States

Academic Press is an imprint of Elsevier
125 London Wall, London EC2Y 5AS, United Kingdom
525 B Street, Suite 1650, San Diego, CA 92101, United States
50 Hampshire Street, 5th Floor, Cambridge, MA 02139, United States
The Boulevard, Langford Lane, Kidlington, Oxford OX5 1GB, United Kingdom

Library of Congress Cataloging-in-Publication Data
A catalog record for this book is available from the Library of Congress

British Library Cataloguing-in-Publication Data
A catalogue record for this book is available from the British Library

ISBN 978-0-323-85413-9

For information on all Academic Press publications
visit our website at https://www.elsevier.com/books-and-journals

Publisher: Mara Conner
Acquisitions Editor: Sonnini R. Yura
Editorial Project Manager: Charlotte Rowley
Production Project Manager: Kamesh Ramajogi
Cover Designer: Victoria Pearson

Typeset by STRAIVE, India

Contents

Contributors

Busayo Oluwatobiloba Aderotoye
Department of Biomedical Engineering, Near East University, Nicosia, Turkish Republic of Northern Cyprus, Turkey

Omar Sameer Adnan
Department of Biomedical Engineering, Near East University, Nicosia, Turkish Republic of Northern Cyprus, Turkey

Mennatullah Ahmed
Department of Biomedical Engineering, Near East University, Nicosia, Turkish Republic of Northern Cyprus, Turkey

Laith M. Alasais
Department of Biomedical Engineering, Near East University, Nicosia, Turkish Republic of Northern Cyprus, Turkey

Hesham Alkahlout
Department of Biomedical Engineering, Near East University, Nicosia, Turkish Republic of Northern Cyprus, Turkey

Rayan Allaia
Department of Biomedical Engineering, Near East University, Nicosia, Turkish Republic of Northern Cyprus, Turkey

Hamza Alloush
Department of Biomedical Engineering, Near East University, Nicosia, Turkish Republic of Northern Cyprus, Turkey

Basel Almagharby
Department of Biomedical Engineering, Near East University, Nicosia, Turkish Republic of Northern Cyprus, Turkey

Abdulrahim S.A. Almoqayad
Department of Biomedical Engineering, Near East University, Nicosia, Turkish Republic of Northern Cyprus, Turkey

Suleyman Asir
Department of Materials Science and Nanotechnology Engineering, Near East University, Nicosia, Turkish Republic of Northern Cyprus, Turkey

Zuhdi Badawi
Department of Biomedical Engineering, Near East University, Nicosia, Turkish Republic of Northern Cyprus, Turkey

Hasan Badran
Department of Biomedical Engineering, Near East University, Nicosia, Turkish Republic of Northern Cyprus, Turkey

Monireh Bakhshpour
Department of Chemistry, Hacettepe University, Ankara, Turkey

Hamdi Burakah
Department of Biomedical Engineering, Near East University, Nicosia, Turkish Republic of Northern Cyprus, Turkey

Adil Denizli
Department of Chemistry, Hacettepe University, Ankara, Turkey

Basil Bartholomew Duwa
Department of Biomedical Engineering, Near East University, Nicosia, Turkish Republic of Northern Cyprus, Turkey

Abdullah Ghader
Department of Biomedical Engineering, Near East University, Nicosia, Turkish Republic of Northern Cyprus, Turkey

Sunsley Tanaka Halimani
Department of Biomedical Engineering, Near East University, Nicosia, Turkish Republic of Northern Cyprus, Turkey

Majed Hejazi
Department of Biomedical Engineering, Near East University, Nicosia, Turkish Republic of Northern Cyprus, Turkey

Mohamad Hejazi
Department of Biomedical Engineering, Near East University, Nicosia, Turkish Republic of Northern Cyprus, Turkey

Kamil A. Ibrahim
Applied Artificial Intelligence Research Centre, Department of Computer Engineering, Near East University, Nicosia, Turkish Republic of Northern Cyprus, Turkey

John Bush Idoko
Applied Artificial Intelligence Research Centre, Department of Computer Engineering, Near East University, Nicosia, Turkish Republic of Northern Cyprus, Turkey

Ahmad Jarwah
Department of Biomedical Engineering, Near East University, Nicosia, Turkish Republic of Northern Cyprus, Turkey

Ahmad Khabbaz
Department of Biomedical Engineering, Near East University, Nicosia, Turkish Republic of Northern Cyprus, Turkey

Kevin Meck
Department of Biomedical Engineering, Near East University, Nicosia, Turkish Republic of Northern Cyprus, Turkey

Mohammed Bin Merdhah
Department of Biomedical Engineering, Near East University, Nicosia, Turkish Republic of Northern Cyprus, Turkey

Sabareela Moro
Department of Biomedical Engineering, Near East University, Nicosia, Turkish Republic of Northern Cyprus, Turkey

Mandy Sizalobuhle Mpofu
Department of Biomedical Engineering, Near East University, Nicosia, Turkish Republic of Northern Cyprus, Turkey

Mustapha Taiwo Mubarak
DESAM Research Institute; Department of Biomedical Engineering, Near East University, Nicosia, Turkish Republic of Northern Cyprus, Turkey

Nyasha T. Muriritirwa
Department of Biomedical Engineering, Near East University, Nicosia, Turkish Republic of Northern Cyprus, Turkey

Rehemah Namatovu
Department of Biomedical Engineering, Near East University, Nicosia, Turkish Republic of Northern Cyprus, Turkey

Dilber Uzun Ozsahin
DESAM Research Institute; Department of Biomedical Engineering, Near East University, Nicosia, Turkish Republic of Northern Cyprus, Turkey; Medical Diagnostic Imaging Department, College of Health Sciences, University of Sharjah, Sharjah, United Arab Emirates

Ilker Ozsahin
DESAM Research Institute; Department of Biomedical Engineering, Near East University, Nicosia, Turkish Republic of Northern Cyprus, Turkey; Brain Health Imaging Institute, Department of Radiology, Weill Cornell Medicine, New York, NY, United States

Ramiz Musallam Salama
Department of Computer Engineering, Near East University, Nicosia, Turkish Republic of Northern Cyprus, Turkey

Jamil Hilal Seif Abu Shaban
Department of Biomedical Engineering, Near East University, Nicosia, Turkish Republic of Northern Cyprus, Turkey

Ismail Ramadan Swalehe
Department of Biomedical Engineering, Near East University, Nicosia, Turkish Republic of Northern Cyprus, Turkey

Serhat Unal
Department of Infectious Disease and Clinical Microbiology, Hacettepe University, Ankara, Turkey

Abdullah A. Usman
Department of Biomedical Engineering, Near East University, Nicosia, Turkish Republic of Northern Cyprus, Turkey

Noman Abdul Wajid
Department of Biomedical Engineering, Near East University, Nicosia, Turkish Republic of Northern Cyprus, Turkey

Majd Zeidan
Department of Biomedical Engineering, Near East University, Nicosia, Turkish Republic of Northern Cyprus, Turkey

CHAPTER ONE

Introduction to biomedical instrumentation

Ilker Ozsahin[a,b,c], Dilber Uzun Ozsahin[a,b,d], and Mustapha Taiwo Mubarak[a,b]
[a]DESAM Research Institute, Near East University, Nicosia, Turkish Republic of Northern Cyprus, Turkey
[b]Department of Biomedical Engineering, Near East University, Nicosia, Turkish Republic of Northern Cyprus, Turkey
[c]Brain Health Imaging Institute, Department of Radiology, Weill Cornell Medicine, New York, NY, United States
[d]Medical Diagnostic Imaging Department, College of Health Sciences, University of Sharjah, Sharjah, United Arab Emirates

Biomedical instrumentation focuses on the development of methods and devices for the treatment of diseases. It is an emerging field of biomedical engineering that bridges the gap between medicine and engineering. Biomedical instrumentation was introduced during the Apollo missions when it became a necessity to measure the vital signs of astronauts. During and after the Apollo mission, biomedical engineers extended the knowledge from the Apollo mission to the research and development of more sophisticated medical equipment that are now used today. Examples include diagnostic equipment (medical imaging devices), durable medical equipment (insulin pumps and kidney machines), therapeutic equipment (infusion pumps, medical lasers, and surgical machines), life support equipment (heart-lung machines, dialysis machine, and incubator), and medical laboratory equipment (chemistry analyzer, blood gas analyzers, electrolyte analyzers, etc.). The application of these devices opens a new phase in the medical industry. Now you can have patients with terminal diseases living longer than usual. This research also extends to the field of artificial organs where vital organs, including the liver, kidneys, and heart, are designed and developed. This present opportunity allows for the effective management of diseases and disorders.

All biomedical equipment and instruments are based on a simple model. This model usually consists of a sensor, an interface, and a computation unit. It is necessary for any tissue or the biological material to interact directly or indirectly with the sensor. This will enable the sensor to monitor and report any physiological changes occurring in and around the tissue. These sensors are biomaterials that are compatible with human tissue and are durable and not harmful to the body. A typical sensor must be capable of providing a safe interface with biological material and detect biophysical, biochemical, and

Modern Practical Healthcare Issues in Biomedical Instrumentation
https://doi.org/10.1016/B978-0-323-85413-9.00005-0

bioelectrical changes or parameters. These parameters include average arterial blood flow, heart rate, or electrocardiogram. The detected signals are further transferred for recognition and interpretation. The electronic interface must be capable of matching the electronic attribute of the sensor. Also, it must be able to preserve the signal-to-noise ratio of the sensor and to provide a protected interface for both the sensor and the computing unit. The computation unit allows for the overall control of the system by a user.

This book details the design, development, and application of biomedical instruments. This includes smart 3D artificial hands, smart-assisted gloves for paralytics, nanoparticle-based plasmic devices, etc. With the rapid growth in biomedical engineering, biomedical instrumentation will continue to explore new ways to improve and understand how the human body works and advance technology to solve problems associated with it.

CHAPTER TWO

Designing a 3D printed artificial hand

Dilber Uzun Ozsahin[a,b,c], Majed Hejazi[b], Omar Sameer Adnan[b], Hamza Alloush[b], Ahmad Khabbaz[b], John Bush Idoko[d], Basil Bartholomew Duwa[b], and Ilker Ozsahin[a,b,e]

[a]DESAM Research Institute, Near East University, Nicosia, Turkish Republic of Northern Cyprus, Turkey
[b]Department of Biomedical Engineering, Near East University, Nicosia, Turkish Republic of Northern Cyprus, Turkey
[c]Medical Diagnostic Imaging Department, College of Health Sciences, University of Sharjah, Sharjah, United Arab Emirates
[d]Applied Artificial Intelligence Research Centre, Department of Computer Engineering, Near East University, Nicosia, Turkish Republic of Northern Cyprus, Turkey
[e]Brain Health Imaging Institute, Department of Radiology, Weill Cornell Medicine, New York, NY, United States

2.1 Introduction

Electromyography (EMG) can be considered as one of the most widely used control interfaces for recent modernized upper limb prosthetics. Prosthetic arms that utilize the EMG interface are referred to as myoelectric arms [1]. The hands that are used in such cases are called myoelectric hands. Subsequently, these technology applications on a commercial scale can be very expensive, as the cost of myoelectric hands ranges from $15,000 to $50,000. Usually, the maintenance procedures needed for these hands demand the presence of expensive proprietary components accompanied by a highly trained and experienced professional technician for repair phases. Nowadays, on a global scale, a lot of amputees cannot benefit from using a myoelectric prosthetic due to the enormous cost involved. The most affordable and commonly used myoelectric hands (almost at $15,000) are known to be lacking much functionality compared to the natural human hand [2]. These hands grant their users one type of grip, which is the pinch grip. The speed of these hands can be adjusted and controlled by the user through the EMG control system used in these hands. However, users must always be careful while trying to hold objects as they might be accidentally crushed by the actuation of the hand. More advanced hands can be found across the world with drastically superior functionality when compared with the most affordable hands. For instance, the BeBionics hands and I-Limb have

Modern Practical Healthcare Issues in Biomedical Instrumentation
https://doi.org/10.1016/B978-0-323-85413-9.00009-8

such features as the ability to switch between many grips and finger movement [3]. Unfortunately, these hands can be extremely expensive for most amputees, putting aside the small percentage of them who might afford it. Because of such a wide gap between affordability and functionality, and with an attempt to make this crucial technology feasible for the common amputee user, we developed a prototype of a hand that has similar features and functionalities to that of the most sophisticated existing myoelectric hands. We achieved our goal of developing a high-tech hand by implementing a user interface that makes it possible for the user to call many hand functions. Accidental movements and unintended actuation of the hand has been reduced by the implementation of the multiple impulse user interfaces that work on reducing the possibility of a stray muscle impulse. This reduces the user's mental task load.

When there is stimulation from the user's residual limb, EMG signals are automatically captured. This signal gets amplified and digitized using an analog to digital converter. Then, the digitized signal gets transferred to a microcontroller by employing a serial peripheral interface. In the microcontroller, the muscle impulses are identified, and the motors in the hand get actuated. The sequence of consecutive impulses produced by the user determines the function performed by the mechanical hand. If a call of a function is detected by control logic of the hand, haptic feedback will alert the user about the status of the device.

2.1.1 Background

The cost of a modern myoelectric prosthetic hand is approximately \$15,000–\$50,000. A price that can be affordable for average arm amputees in developed countries is, however, not affordable for the average arm amputees in the developing countries. The high price of myoelectric hands makes it cumbersome for amputees all over the world. In addition, the hand functionality and features are limited among the cheaper priced devices. For instance, many of these devices have only one feature, which is to open and close in a single grip. Moreover, upper limb prosthetics had great advances in terms of functionality, especially for the high-end products. For example, products like BeBionic and I-Limb hands on the market operate by using independently actuated fingers and have many options available when it comes to grips. Unfortunately, the high costs linked with such devices prevent most amputees from purchasing it. Thus, when a search is made in the market for these products, scant numbers are available, and they are relatively expensive to purchase. This study builds upon this limitation by developing a myoelectric hand with the same features and functionality as the most expensive ones at a lower cost. Two parallel systems were developed: a low-cost electromechanical hand that has a sensible analog of the human hand and a low-cost EMG-based control platform. The mechanical hand components were printed by a 3D printer in order to lower the cost and to catalyze the manufacturing process. Also, the EMG control platform is comprised of low-cost, ready-to-use, and available components. Since 1986, 3D printing has been around and available. This technology has not gone viral due to the fact that it is expensive, has high-cost of processing, and produces fragile parts. In the past few years, these paradigms shifted greatly. Many tech industries introduced high quality consumer 3D printing devices, such as the Bits made by RapMan and the Makerbot Replicator. This facilitates the use of this technology for a large number of users. Nowadays there are more than 24 consumer-grade 3D printers with a price under $5000. Most of these printers depend on the usage of the fused deposition modeling (FDM) as a printing technique, which can provide the most complicated and robust parts of any 3D printing technology [1].

2.1.2 Models

EMG signal acquisition systems have components, such as the analog filters and gain stages, for every channel. In order to achieve the goal of limiting the use of components to the minimum for the proposed system to have fewer components than that of a traditional EMG acquisition system, we employed a different approach to this problem. A high resolution analog-to-digital converter is used for capturing the signal with a resolution that is of a few hundred nanovolts. This resolution level has the ability to perform filtering on the signal in the digital domain. The best characteristic of the microcontroller is its ability to perform all types of signal processing. Fortunately, microcontrollers have become more affordable and widely available in the past few years. Arduino microcontrollers have been ranked as one of the most affordable and available microcontrollers on the market. They are characterized by being affordable and also with their simple C++ based programming environment. However, there is constant advancement, and more powerful microcontrollers have been produced. The Due, the fastest available Arduino microcontroller, runs at a clock frequency of 84 MHz. But there is more speed in the Beagle bone, a new microcontroller that is almost the same size as the Arduino, that can operate at up to 700 MHz. The increase of quality and access to manufacturing techniques and advanced components makes it possible for a lot of older, high-cost specialized applications to be available at a much lower cost for the public. This is already happening in 3D printing and will be witnessed in 3D-printed prosthetic hand studies.

2.1.3 Prosthetic hand overview

The primary components for the low-cost prosthetic hand developed under this study are an EMG interface, a 3D-printed electromechanical hand, a stable embedded control system, and a microcontroller able to do real-time processing for signals. The 3D-printed hand prototype cost less than $250 for modeling, printing, and assembling. The hand has more than 30 components that include 15 unique printed parts. High-torque hobby servos actuate it. These servos are controlled by pulse with modulated signals (PWM). The Arduino Due microcontrollers regulate these signals.

The EMG interface system functions through taking differential signals from impulses of the muscles in the user's residual limb. The acquired signals are then amplified and passed to a high resolution analog-to-digital converter (ADC). The ADC is responsible afterwards for outputting the signals over a serial peripheral interface (SPI) to the Arduino Due microcontroller.

The Arduino processes the acquired signals by performing a Fast Fourier Transform (FFT) that converts the signals from time domain to frequency domain. Once a signal reaches the frequency domain, the magnitude of its relevant frequency bins is calculated. A muscle impulse is detected on that channel if the magnitude of those relevant bins becomes larger than the threshold value. Combinations of muscle impulses in the available channels are captured by the control logic embedded in the Arduino Due. These combinations are called opcodes. The sequences of opcodes is analyzed by the control logic that actuates the motors of the hand to perform the functions dictated by the given opcode sequence. This method of using opcode sequence to dictate the functions of the hand is called Multiple Impulse User Interface (MIUI).

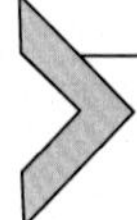

2.2 Health and safety

2.2.1 Factors of safety

In this study, three primary factors were considered when assessing the electrical safety of medical devices: the duration of current flow, the pathway it takes, and the magnitude of the potential current. A myoelectric arm has a large battery that can drive 3 amp of DC current in order to power the motors at the needed maximum torque. This current is considered more than sufficient to cause many severe problems such as respiratory paralysis, fatigue, and pain. It might also lead to triggering a sustained myocardial contraction. The beating of the heart stops while contracting; however, a normal beating is restored when the current supply stops and is no longer applied. So, as current is not sustained for a long time, lethality is prevented. The possibility of a 3 A discharge into the body is minimal, but on the other hand, much less current than 3 A, when applied to a vulnerable body part, would be of extreme danger to it.

The path of the current has an eminent effect on how serious the electric shock would be. When current passes through the cardiac tissue of a person, a microshock occurs. These are more likely to be lethal than macroshocks, even with a relatively small current as 10 μA. The heart requires huge magnitude of current for fibrillation to occur, especially when it is applied through the skin. To avert potential harm resulting from small leakage currents that passes through cardiac tissue, it is crucial that specific precautions be taken to prevent such leakage currents in devices. The main solution for limiting electrical risk in medical devices by using an isolated power supply system. In this iteration, proper isolation was not provided; however in the next iteration, a capacitive or an optical isolation stage must be implemented. This isolation stage should be positioned between the microcontroller and on the residual limb of the users. This way, the exposure of possible current on the user is limited in the event of any electrical discharge into the body.

2.3 Fundamentals of surface electromyography

The basis of the skeletal muscle organization is the motor unit (MU). Every MU is made of motor nerve fibers that are connected to a group of muscle fibers. An MU is the tiniest muscle element voluntarily activated in this research. The activation of one MU shows itself as a distributed bipotential because of the superposition of action potentials of every muscle fiber that is in the bundle. The amplitude of an EMG signal (myoelectric signal) is measured in the range of 20 and 2000 μV, depending on the size of a MU. When recording EMG signals, the frequencies of interest are in the range of 30–300 Hz. As seen in Fig. 2.1, it is possible for the contraction of the muscle to occur in a distinctive activation of one MU, however, on the condition of employing intramuscular probes and correct signal filtering.

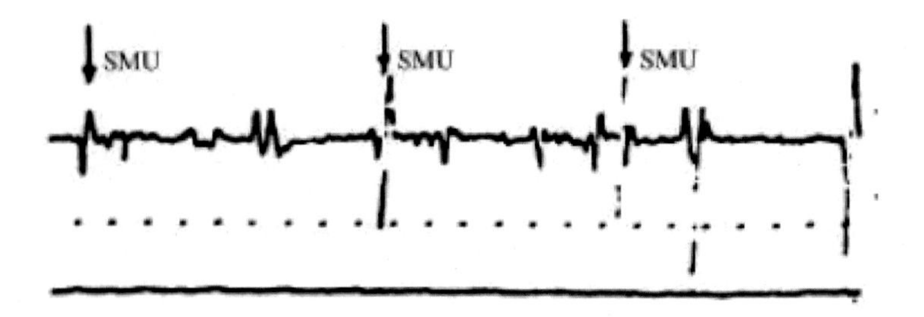

Fig. 2.1 Clear visibility of activation of single motor unit.

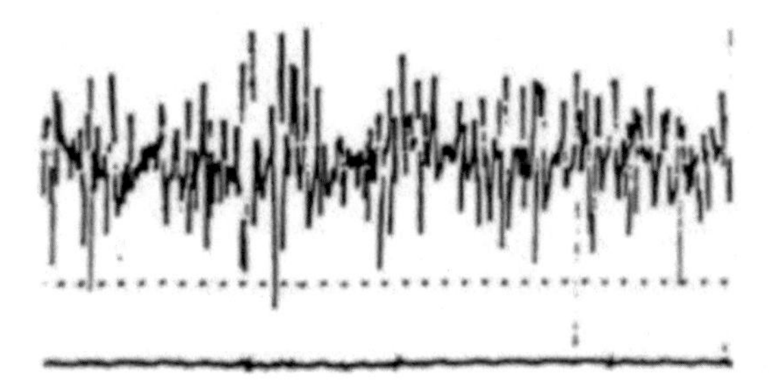

Fig. 2.2 Extreme noise appearing in multiple motor unit activation.

When the contraction force value increases, it is impossible to distinguish each individual MU in the time domain signal. This combination is due to increased MU recruitment (increase in MUs firing) as well as an increase in activation rate of MU. The result is illustrated below in Fig. 2.2, a time domain similar to stochastic noise.

The difficulty of isolating the EMG signal from a single muscle increases due to the noise-distributed signal [4]. The problem gets worse as a surface EMG (SEMG) is employed. SEMG demands a large superficial muscle (pronounced muscle group) unlike intramuscular, which is needle-based myoelectric signal acquisition. It is required to locate closely spaced electrode pairs positioned along the fiber of the muscle to acquire a highly localized signal that can be fit for a single muscle control channel. In an attempt to locate these electrode pairs, the recorded signal has to be spatial and temporal, and should superposed the electrical activity for all the active MUs in the muscle group.

Electrical noise makes it more complicated to interpret these complex signals. The main source of electrical noise in almost all bioinstrumentation applications comes from capacitive coupling between the power delivery lines and the human body. Mainly noise occurs when the frequency is harmonic.

2.3.1 Application of electromyography in prosthetics

Myoelectric hands are considered as the type of artificial hands that have their control system based on electromyography [5]. Purposeful muscle contractions are the controllers of the myoelectric prosthetic devices, and usually, these contractions are in the user's residual limb. Electrodes are held against muscles or groups of muscles in the residual limb of the user by tightly fitting prosthetic sockets.

The role of these electrodes is to register the surface EMG signals. Myoelectric arm users can open and close myoelectric hands and change between the functions of the hands by carefully controlling the contractions of the muscles [6]. Fig. 2.3 shows a clear demonstration of how the control loop is employed in a myoelectric arm.

Control systems in myoelectric hands can be in single- and multichannel configurations. Most times, the fitting of single-channel control systems is used for children or people who have fewer muscles in their residual limb. On the other hand, two-channel control systems are mostly used.

It is impractical to use a single muscle per control channel on an electrode pair, which is necessary over the muscle to be observed. While dealing with muscles and group signal acquisition, it is very important to consider the exact location of the electrode placements since the prosthetic sockets might move or shift while in use. Due to this reason, it is very common for muscle group signal acquisition to be used as a control approach. Again, it is easier to obtain and process more complex EMG signals than trying to maintain single muscle channel control of prosthetic products. Features that show the intention of the user from these complex signals must be extracted in order to control the myoelectric device. Most of the available devices on the market

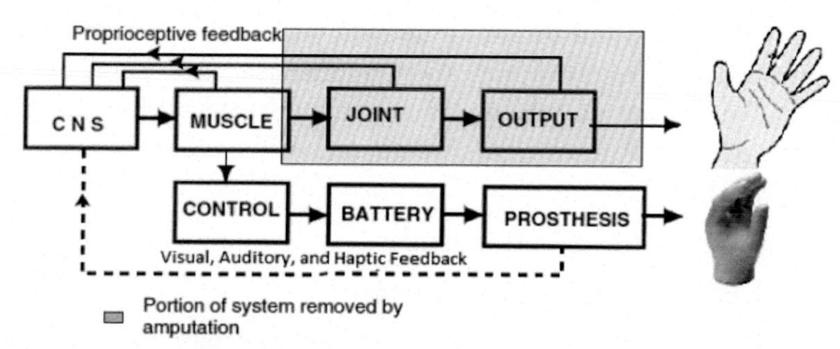

Fig. 2.3 Control loop in myoelectric arm.

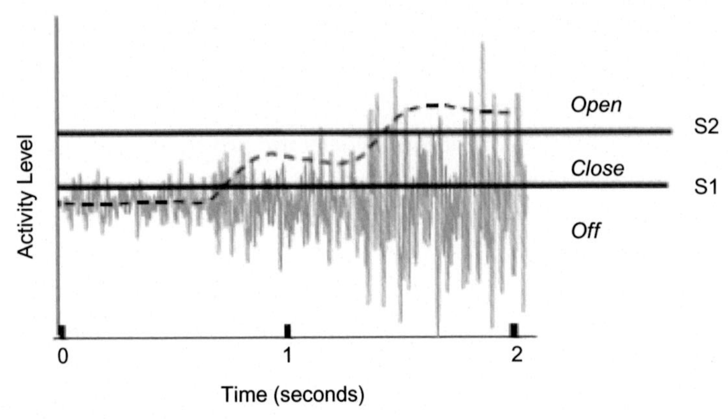

Fig. 2.4 Amplitude control for single channel.

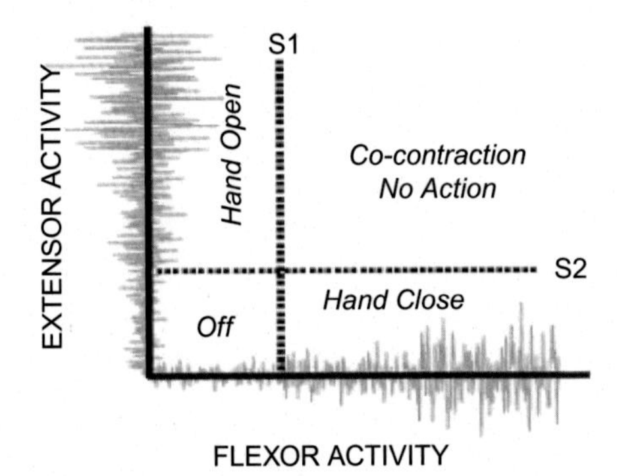

Fig. 2.5 Amplitude control for dual channel.

use signal amplitude to evoke a control system response. This control method is commonly applied in single- and multichannel devices as depicted in Figs. 2.4 and 2.5 respectively.

In the past, pattern recognition control systems have been used to work on multifunction control systems. These systems lack intuitiveness and basic functionality and features of most of the limbs available today, and for these reasons, they are not normally used. A simple control system is required for a limb that has only two functions.

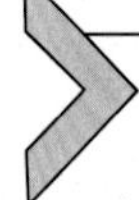

2.4 Items used in the study

2.4.1 MyoWare muscle sensor

Object controlling by muscles is an interesting problem in the engineering field. A kit called the MyoWare muscle sensor (Fig. 2.6) manufactured using

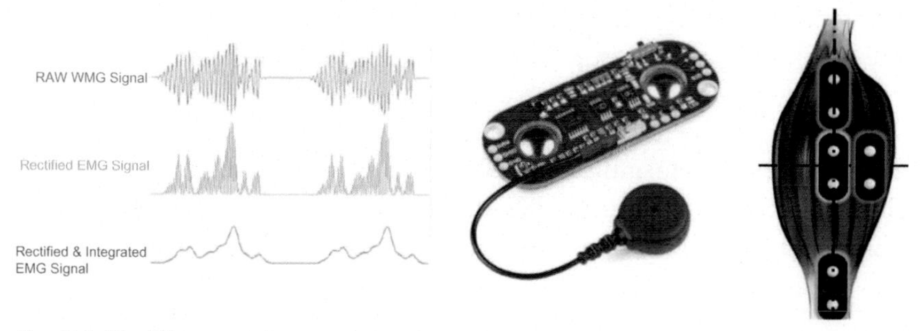

Fig. 2.6 MyoWare muscle sensor.

advanced technology has the ability to perform the main function of the sensor; to measure the electrical activity of the muscle; and also to rectify, amplify, and filter them to produce analog output signals, which can be easily read by an Arduino microcontroller [7]. What makes it really unique is that it is easily wearable; cleaning the area of the muscle is an important step to get rid of dirt, which improves the conductivity when sensing the signal from the muscles, then snapping three electrodes to the sensor. The first snap goes to the mid-muscle electrode, the second snap to the end muscle electrode, and the last snap to the reference electrode, then placing the sensor on the muscle that the user willingly targets, taking into consideration the right place of each electrode so that the signal can be read easily by the sensor. The first electrode is placed in the middle of the muscle while the second one is placed at the end of the muscle, and the third one is placed on a bony part, such as the elbow, not far away from the muscle. The right position of the electrodes is an important factor that determines the strength of the given signal. The sensor is supplied by a 3.7 V lithium battery; the power supply (+Vs) is connected to the positive side of the battery. The power supply ground is connected to the negative side of the battery, and the output signal is connected to the analog input A0 of the Arduino.

2.4.2 Servomotors

Servomotors are devices/actuators that are used in many applications such as robot cars, boat cars, etc. They are available in many sizes either in small sizes for small projects or in big sizes for industrial purposes. This study contains five servomotors, each finger connected to a single motor where the motors move freely [8]. The movement of one finger does not depend on the movement of the other fingers. To understand what they are or how they work, it is paramount to know the inner most content. They contain a simple setup circuit and a motor potentiometer; the motor is controlled by gears, and as the motor starts to rotate, the value of the resistor changes, allowing the control circuit to control the direction and the movement of the servomotor. The movement of the servo is controlled by the Arduino microcontroller or pulse with modulation of the Arduino microcontroller. And its main function is to determine the direction of the rotation, either clockwise or counter clockwise, based on the duration and the pulse, which is sent by the control wire.

2.4.3 Step-down regulator

A step-down regulator is a power converter that does nothing with AC or DC-to-DC converter. The main function of the step-down is that it triggers the input voltage to step up the current in the main parts of the two semiconductors, and one storage element capacitor or inductor also can be the combination of both of them, which does the smoothing or reducing the voltage Ripple.

2.4.4 Rechargeable lithium batteries

The rechargeable lithium batteries are regarded as the best storage devices in the world. They play important roles in the portable electronics sector and are the only solution in the sector of automotive. What makes them the best in the market is their efficiency. Over 95% of their lifecycle is of 3000 cycles as well as their energy density and power density. In this study, lithium ion batteries 1 and 2 are connected in parallel with each other and are called A battery. Another 3 and 4 are connected in parallel as well and are called A and B, which are also connected in series. Each battery is 3.7 V, so performing some calculations, they are giving out 7.4 V, which is too much for the servomotors to withstand; that is why we used a step-down voltage regulator.

2.4.5 Arduino and pulse with modulation outputs

Arduino Uno is a board that consists of 14 digital output/input pins, 6 of which are PWM (pulse with modulation) outputs, which describe a digital signal type. Pulse with modulation can be used in many different applications, such as controlling the brightness of light-emitting diodes (LED) or controlling the direction or speed of servomotors; in other words, PWM allows the users to control different electronic devices [9]. The power supply is pulsed on and off at a specific frequency and with a certain pulse width in general. The main function of pulse with modulation outputs is that they enable efficient power supplies, such as step-up and step-down, and regulate power supplies.

Duty cycle/50% duty cycle occurs when half of the digital signal is ON and when half of it is OFF. Similarly, when 75% of the cycle is ON and 25% is off, it is called 75% duty cycle. The Arduino microcontroller also has six analog inputs (six channels) or pins of analog-to-digital converters. The function of these channels, or these pins, is that they can easily read the analog sensors since the sensors take or calculate different values. Each one of the 6 analog inputs provides ten bits of resolution (1024 different values). There's also a USB connection in the Arduino microcontroller, either to get the required power needed by the Arduino or upload the written code to it [10]. A power jack to power the Arduino by an external power source, usually a 9-V battery, is also used. The Arduino board also contains 8 power pins, which is sometimes labelled as 9 V; this is the input to the Arduino when powering it by an external power source. Also, there are power part 5 V, 3.3 V supplies in addition to the ground pins. This implies that the board has everything required to support the microcontroller.

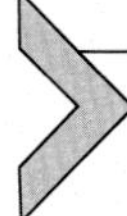

2.5 Method

2.5.1 Hand design

The finger drawings of the hand were done on the SolidWorks program. The fingers' connections were the most important aspect of the artificial finger design. The relationship among those connections creates movement freedom, or ease of movement. Connecting pins, parts of palm, part of wrist, part of arm, and motor connection parts forms the drawings. The hand design is carefully drawn as the fingers contain very thin tubes or rope line that connect the fingers to the servomotors, and all these parts were printed by a 3D printer [11].

2.5.2 Connection of circuit

As we see from the block diagram (Fig. 2.7) and the circuit above the 3D printed hand, the first component is the lithium batteries [12]. Each battery is 4.2 at its maximum voltage value. Here, 4 batteries were connected in both parallel and series with a total voltage of 8.4 V, but each servo needs a range of voltage between 3.7 and 6, requiring a step-down regulator so that the servomotors can receive 5 V. The batteries and the servomotors

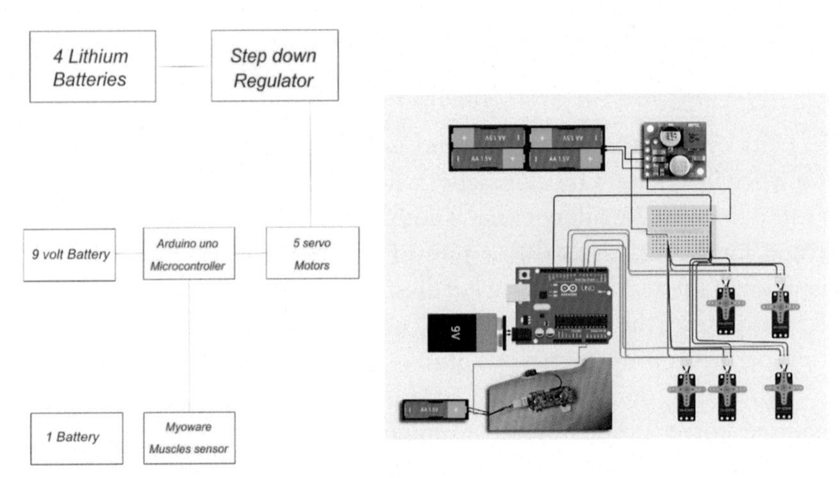

Fig. 2.7 Block diagram and study simulation.

are connected to the pulse and modulation outputs of the Arduino of the pins number 1, number 3, number 5, number 6, number 9 and number 10. As known, servomotors have three wires: orange (PWM), brown (ground), and red (positive). Ground and positive wires are connected to the breadboard; the negative and positive 5-V output terminals of the step-down regulator are connected to the breadboard. In other words, the output wires of the step-down regulator are connected to positive and ground terminals of the servomotors. The wire connected to the ground of the Arduino should also be connected to the negative terminals of the servomotors through the breadboard. Arduino is powered by an external power source, which is a 9-V battery. Finally, the MyoWare muscle sensor that is placed on the target muscle senses the signal from muscle and then sends it to the analog input of the Arduino microcontroller. The sensor is powered by a 4.2V lithium battery and is connected to the Arduino.

2.6 Result and discussion

To regulate the operation of movement detection, an EMG sensor obtains signals from the targeted muscles; the signals that control the arm are received and processed by the microcontroller, which then transmit the signals to the servomotors. The signal taken from the group muscles is tested, and the best result is determined and further accepted. Different conducted studies have proven that electrodes and EMG sensors generate different signal values on participants

The proposed 3D-printed prosthetic hand closes and grabs objects when A0 signal values are above 600 or 3.3 V, so the hand performs as desired [13]. Signals can be acquired either from the biceps or the forearm. The movement of each finger doesn't depend on the movement of the other fingers. The purpose of this study is to make a cheap and practical prosthetic hand for patients who lost their hands due to specific reasons such as birth defect or accidents and also for those who cannot afford an expensive traditional prosthetic hand.

The effectiveness of the 3D-printed myoelectric hand was tested by grasping a variety of different objects [14]. The device has the ability to grasp objects such as a towel, a purse, a can of food, smart phones, juice bottles, etc. And this shows that the purpose of this study is achieved. The implemented system has shown practical tendencies of having the best qualities compared to other products.

2.7 Comparison

This section illustrates the comparison between the 3D-printed prosthetic hand controlled by a MyoWare muscle sensor and myoelectric prosthesis (MP) as depicted in Tables 2.1 and 2.2. Despite advancement in technology, the 3D-printed hand is still not widely used. The proposed 3D-printed hand has showed weaker gripping strength compared to the

Table 2.1 Comparison between myoelectric and 3D printed prosthetic hand with EMG sensor.

	Myoelectric prosthesis	3D-printed prosthetic hand
Weight	840 g	700 g
Sensor	2 Myoelectric sensors	1 MyoWare muscle sensor
Battery	Li-ion 7.2 V (65 g)	Li-ion 4.2 V (60 g)
Cost	$13,000 USD	$800 USD
Manufacturing time	14 days	4 days

Table 2.2 Comparison between the activities of each hand.

	Myoelectric prosthesis	3D-printed prosthetic hand
Eating snacks	Yes	No
Dressing the button	Yes	No
Writing	Yes	No
Wearing the jacket	Yes	Yes
Drinking tea from the cup	Yes	Yes

other myoelectric hand. It is known that the technology of 3D printing faces many limitations. Nowadays, instead of replacing the prosthesis, there are numerous advantages such as short manufacturing period, low cost, and weight. The 3D-printed hand can improve the quality of life for amputees, especially those who are uninsured or who face economic hardships. Furthermore, amputees can print a new hand in an easy way as required with a relatively low cost. Additionally, if technology of 3D printing is partially introduced to the fabrication of prosthesis, a higher quality of the prosthesis and lower cost can be expected. Also 3D printing technology is used in many different fields such as engineering, medicine, etc. For example, generation and transplantation of tissues are done by 3D bioprinting technology [15]. This 3D printing technology is considered sensitive as it can respond to individuals' needs and demands as materials used can be changed in an easy way. Prosthesis fabrication applications of 3D printing technology for rehabilitation of people who are disabled can overcome the burden of the cost and the amount of time needed for conventional prosthesis manufacturing. On the other hand, the proposed system has different limitations including the short life of lithium ion batteries as well as the noise produced by the servomotors during the motion. Secondly, the patient repulses the hand due to the strong mechanical feeling that happens as a result of the continued use of the hand. The third is the weak gripping strength of the hand; here the mechanical components are not durable and can be damaged easily by shock. The performance of MP is compared with the results achieved by the proposed system. The weight of the MP is around 800 g. It can be operated by using two sensors, called myoelectric sensors, that detect the electrical activity of both Extensor Carpi Radialis (ECR) and Flexor Carpi Radialis (FCR) muscles. The hand was manufactured to sense the voluntary opening motion from extensor movements with an attached electrode by the application of electric signals and to sense the voluntary closing with flexor movement. The hand is made of silicon, and it is considered a cosmetic prosthesis. The cost is around $13,000 (USD). Therapy training should be carried out every day for 30 min for the patient to get used to the hand.

2.8 Conclusion

In this research, it is clear that it is possible to create a 3D-printed prosthetic hand with no huge cost and effort but with a satisfying result. The mechanical design of this study was simple and successful, and most importantly, this hand can be printed by anyone, which is a big advantage.

Similarly, its design can be modified any time to suit the size and shape of the patient's body. The materials used in the printed hand can always be changed to improve the performance. Carbon fiber, which is a light and hard material, can be used. Pressures or sensors can also be added to the fingers to give a groping strength of the hand when groping something. The explored MyoWare muscle sensor gives the device the capabilities to get and sense the electrical signals from the muscle fibers, enabling an accurate control of the hand. Varieties of techniques were employed to get battery sensing of the signal, either by using more than one sensor for separate control or by using an invasive technique called targeted muscle inversion. As previously mentioned, it is invasive because it requires a surgical procedure. Arduino is also used to translate the signal from the target muscle into a movement to break the specified threshold intensity, which causes the servomotors to rotate according to the angle specified in the code.

References

[1] T. Zhang, H. Liu, L. Jiang, S. Fan, J. Yang, Development of flexible 3-D tactile sensor system for anthropomorphic artificial hand, IEEE Sensors J. 13 (2) (2012) 510–518.
[2] L. Wu, M.J. de Andrade, L.K. Saharan, R.S. Rome, R.H. Baughman, Y. Tadesse, Compact and low-cost humanoid hand powered by nylon artificial muscles, Bioinspir. Biomim. 12 (2) (2017), 026004.
[3] Z. Xu, E. Todorov, Design of a highly biomimetic anthropomorphic robotic hand towards artificial limb regeneration, in: *2016 IEEE International Conference on Robotics and Automation (ICRA)*, IEEE, 2016, May, pp. 3485–3492.
[4] R. Schuster, F. Paternoster, W. Seiberl, High-density electromyographic assessment of stretch reflex activity during drop jumps from varying drop heights, J. Electromyogr. Kinesiol. 50 (2020) 102375.
[5] M. Hahzaib, S. Shakil, Hand electromyography circuit and signals classification using artificial neural network, in: 2018 14th International Conference on Emerging Technologies (ICET), IEEE, 2018, November, pp. 1–6.
[6] E. Elme, M. Larrier, C. Kracinovich, D. Renshaw, K. Troy, M. Popovic, Design of a biologically accurate prosthetic hand, in: 2017 International Symposium on Wearable Robotics and Rehabilitation (WeRob), IEEE, 2017, November, pp. 1–2.
[7] R.E. Russo, J.G. Fernández, R.R. Rivera, Algorithm of myoelectric signals processing for the control of prosthetic robotic hands, J. Comput. Sci. Technol. 18 (1) (2018) e04.
[8] Y. Elmaati, L. El Bahir, K. Faitah, The rotor and tower vibrations damping monitoring in the case of total pitch servomotors failure, Int. J. Acoust. Vib. 25 (1) (2020) 62–72.
[9] M. Abdulhamid, K. Njoroge, Irrigation system based on Arduino uno microcontroller, Poljopr. teh. 45 (2) (2020) 67–78. *J. Prosthet. Orthot.* 2020. Prosthetic 3D Printing. 32, pp. 35-39.
[10] J. Son, W. Zev Rymer, Longer electromechanical delay in paretic triceps surae muscles during voluntary isometric plantarflexion torque generation in chronic hemispheric stroke survivors, J. Electromyogr. Kinesiol 56 (2021) 102475, https://doi.org/10.1016/j.jelekin.2020.102475.
[11] L. Saharan, A. Sharma, M.J. de Andrade, R.H. Baughman, Y. Tadesse, Design of a 3D printed lightweight orthotic device based on twisted and coiled polymer muscle: iGrab

hand orthosis, in: *Active and Passive Smart Structures and Integrated Systems 2017* (Vol. 10164, p. 1016428), International Society for Optics and Photonics, 2017, April.
[12] P. Manikandan, M. Parekh, V. Pol, Enabling longer-lasting rechargeable batteries via synchronizing lithium and lithium-Ion batteries, in: *ECS Meeting Abstracts*, MA2020-01 (2), 2020, p. 452.
[13] R. Geizans, Developing 3D Printed Prosthetic Hand Model Controlled by EMG Signal from Forearm, Metropolia University of Applied Sciences, Helsinki, Finland, 2018.
[14] M. Geoffroy, J. Gardan, J. Goodnough, J. Mattie, Cranial remodeling orthosis for infantile plagiocephaly created through a 3D Scan, topological optimization, and 3D printing process, J. Prosthet. Orthot. 30 (4) (2018) 247–258, https://doi.org/10.1097/JPO.0000000000000190.
[15] J. Koprnický, P. Najman, J. Šafka, 3D printed bionic prosthetic hands, in: 2017 IEEE International Workshop of Electronics, Control, Measurement, Signals and their Application to Mechatronics (ECMSM), IEEE, 2017, May, pp. 1–6.

CHAPTER THREE

Construction of smart assistive gloves for paralytic people

Dilber Uzun Ozsahin[a,b,c], John Bush Idoko[d], Ahmad Jarwah[b], Hasan Badran[b], Noman Abdul Wajid[b], and Ilker Ozsahin[a,b,e]

[a]DESAM Research Institute, Near East University, Nicosia, Turkish Republic of Northern Cyprus, Turkey
[b]Department of Biomedical Engineering, Near East University, Nicosia, Turkish Republic of Northern Cyprus, Turkey
[c]Medical Diagnostic Imaging Department, College of Health Sciences, University of Sharjah, Sharjah, United Arab Emirates
[d]Applied Artificial Intelligence Research Centre, Department of Computer Engineering, Near East University, Nicosia, Turkish Republic of Northern Cyprus, Turkey
[e]Brain Health Imaging Institute, Department of Radiology, Weill Cornell Medicine, New York, NY, United States

3.1 Introduction

One of the greatest abilities of the human body is the movement. Effortless movement can make one's life much easier and comfortable. Some people lose their ability to move because of some diseases or some disorders in their system. This situation is called paralysis, and it is the loss of the ability to move one or more of the human muscles. Paralytic patients suffer a lot in their daily lives, hence doctors and engineers are constantly trying to make these patients' lives easier and more comfortable.

The aim of this study is to build a device that is capable of reducing the communication gap between the paralytic patient and their caretakers, whether they are staying either in the hospital or in homecare. A device would alert the patient's family or the hospital staff of the patient's needs at any given time. The device consists of a couple of Arduino Unos and an accelerometer sensor that observes the movement or gesture made by patients according to what they desire [1]. For instance, if a patient needs food, he will move his hand slightly to the right, and then, the accelerometer will observe this movement and convert it to signals that will be sent using the nRf sensor. These signals will be received by another nRf sensor connected to an Arduino Uno at the other end. The light-emitting diodes (LEDs) on that Arduino Uno are then turned on according to the movement of the hand gesture demonstrated by the patient. Every color will represent a specific message. For instance, the color green indicates that the patient

Modern Practical Healthcare Issues in Biomedical Instrumentation
https://doi.org/10.1016/B978-0-323-85413-9.00004-9

needs water. In line with this, the patient's needs are fulfilled quickly, and the patient's care system will improve.

3.1.1 Causes of paralysis

Paralysis is the loss of the ability to move at least one muscle. It is often caused by damage in the nervous system, especially in the spinal cord. If the spinal cord is damaged through a spinal cord injury, it can cause a disruption of signals to other areas of the body and results in paralysis that can be complete or partial depending on the type of paralysis. There are also many other causes for paralysis [2].

Trauma: Trauma is defined as severe emotional shock and pain caused by an extremely upsetting experience. Trauma can be caused due to a devastating negative impact causing lasting impact on a person's emotional stability and mental level. Many other causes of trauma can be physically brutal in nature while others can be psychological [3].

Nerve injury: Different types of information are carried by nerves; these include information related to electrical impulses, impulses from sensory receptors, impulses from the brain, etc. Once this system goes to a state of disorder, a patient can become paralyzed. In most of the cases, paralysis is caused by nerve damage [4]. Nerve damage due to injury or disease cases leads to paralysis (permanent or temporary) because these injuries affect the person's ability to move and feel [5]. As nerves in our body carry different information to different parts of the body, whenever these nerves get damaged, they are no longer able to transmit the signals.

Poliomyelitis: Poliomyelitis is an ailment that is caused by poliovirus. This virus severely affects the spinal cord, which becomes the main reason of muscle weakness and might lead to paralysis. The main threat of this virus is that it is irreversible. This virus can enter into an environment through an affected person's feces and then spread widely into the community.

Cerebral palsy: This source of paralysis is composed of two words: Cerebral (related to the brain) and Palsy (referred to as weakness). Cerebral palsy is loosely translated as brain paralysis [6]. This source of paralysis is caused due to the damage of part(s) of the brain that control(s) motor activity (movement of body). Cerebral palsy is a common disability that does not get worse with time, but it does get better with time and is treatable [7].

Peripheral neuropathy: Peripheral neuropathy is caused by harm to the fringe nerves. This causes severe pain and can affect different areas of the body. This can cause paralysis if the nerves are damaged completely, but total paralysis is a rare case in neuropathy [8]. Depending on the causes of peripheral neuropathy, it lessens with proper treatment as peripheral nerves have an impressive ability to heal on their own [9].

3.1.2 Types of paralysis

There are numerous sorts of paralysis in light of the fact that there are incalculable ways that the body can be harmed [10]. There are four primary classes of loss of motion that deal with the segment of the body that is influenced. The names are herein listed.

Monoplegia: Monoplegia is the loss of movement for a solitary district of the body, most ordinarily one appendage. Regardless of its nature, the cerebral loss of motion is the principal wellspring of monoplegia. Various different wounds and infirmities can prompt this type of fractional loss of motion, including strokes, tumors, motor neuron damage, brain wounds, and impacted or disjoined nerves at the influenced area [11].

Strokes: A stroke is a remedial emergency that requires immediate treatment. Side effects tend to happen unexpectedly yet will vary depending on the bit of the cerebrum that is affected. Some symptoms of a stroke include severe cerebral pain, impairment or loss of vision, confusion, loss of motion, or shortcoming of an arm, leg or side of the face. Types of strokes include ischemic strokes and hemorrhagic strokes [12, 13].

Hemiplegia: Hemiplegia is a type of loss of motion that impacts just a single side of the body. A related condition, hemiparesis, is a vital loss of value and adaptability on one side of the body, without full loss of movement. Some individuals with hemiplegia develop the condition after a scene of hemiparesis. Others may switch forward and backward between times of hemiparesis and hemiplegia.

The brain is isolated into two hemispheres, isolated by a heap of filaments called the corpus callosum. As a rule, the right side of the brain controls muscles and different capacities on the left half of the body, while the left half of the brain controls a great part of the right side of the body. Hence, hemiplegia and hemiparesis quite often demonstrate an issue with one side of the brain [14].

Hemiplegia may occur suddenly, or grow gradually after some time. A condition identified with hemiplegia, spastic hemiplegia, makes the muscles stall out in a constriction, bringing about little muscle control, perpetual muscle torment, and flighty developments.

Paraplegia: Paraplegia can impact the two legs and the hips. People with this condition cannot walk, move their legs, or feel anything underneath the waist. Spinal cord injuries are the most widely recognized reason for paraplegia. These wounds block the mind's capacity to send and get motions beneath the site of the damage [15]. Some causes include spinal cord infections, spinal cord lesions, and brain tumors.

Quadriplegia: Quadriplegia, which is as often as possible alluded to as tetraplegia, is a loss of movement underneath the neck. In some cases, quadriplegia is a fleeting condition as a result of cerebrum wounds, stroke, or brief weight of spinal string nerves. Spinal string wounds are the principal wellspring of quadriplegia. The most notable explanations behind spinal cord wounds include vehicle crashes, falls, sport injuries, etc.

3.2 Materials and components

In this study, we used the Arduino Uno to complete the connections and to set a program to it to achieve the set goal. Some specific sensors were used in this study to solve the proposed problem. These sensors are the nRf sensor and the accelerometer sensor; we also used batteries to supply power to the breadboard used to connect the LEDs and buzzer to the Arduino. The following section discusses the materials that were used in this study.

Arduino Uno: The Arduino Uno is a microcontroller board. It is an open-source electronics platform based on easy-to-use hardware and software. It is used to read the input inserted into it, such as light sensors, motion sensors, fingerprints, and others. After reading these inputs, it turns them into outputs, including turning the LEDs on or switching motors, etc. One can use the board for any purpose by setting instructions into the microcontroller, which is located in the board. We also used some high-level language codes and the Arduino software.

nRf sensors: In this study, two types of nRF sensors were used: the transmitter module and the receiver module [16]. The first type of sensor is the RF transmitter module, a tiny printed circuit board (PCB) subassembly responsible for transmitting a radio signal and holding data by modulating the wave. Usually, transmitter modules are applied alongside a regulator that supplies the module with data that can be transmitted. RF transmitters are normally subject to control criteria that include the maximum allowable power output of the transmitter, harmonics, and band edge requirements.

The other type is the RF receiver module, which is used to obtain and demodulate the modulated radio wave. There are two kinds of modules for RF receivers: superheterodyne and superregenerative. Using a range of amplifiers to collect modulated data from a carrier wave, superregenerative modules are typically low-cost and low-power designs, but because their operating frequency varies significantly with temperature and power supply voltage, superregenerative modules are usually unreliable. Superheterodyne receivers do have significant advantage over superregenerative ones; over a wide voltage and temperature spectrum, they provide improved precision and stability. This stability stems from a fixed crystal nature that has translated to a slightly more costly product in the past.

Accelerometer sensor: An accelerometer is an electromechanical instrument used for the calculation of the forces of acceleration [17]. Like the relentless force of gravity pushing one's feet, these forces may be static, or they may be dynamic, induced by the accelerometer shifting or vibrating. The mass compresses the crystal within the instrument when the accelerometer is exposed to an accelerating force, causing it to generate an electrical signal. Using built-in electronics that produce an output signal, the signal is then amplified and conditioned, which is ideal for use by a controller or higher level data achievement. In the analysis, the use of the accelerometer is to detect the movement of patient gloves and translate it to some messages sent to an Arduino program to alert the patient's caregiver to attend to the patient's needs.

Breadboard: A breadboard is widely used in the method of building a circuit. When using a breadboard, you do not need to connect wires and components to create a circuit. It is easier to base and repurpose parts. Because components are not joined, you may adjust your circuit design at any point without any disruption. It consists of a series of conductive metal clips wrapped in a white acrylonitrile butadiene styrene (ABS) plastic case, where each clip is secured by another clip. There are a variety of openings, arranged in a particular way, in the plastic case. Two types of areas, also called strips, are included in a standard breadboard layout: bus strips and socket strips. Bus strips are typically used to provide the circuit with power. They consist of two columns, one for the ground and one for the power voltage.

To keep most of the components in a circuit, socket strips are used. They are usually made up of two parts, each with 5 rows and 64 columns. From the inside, every column is connected electrically.

LEDs: Light emitting diodes (LEDs) are used to determine if the messages are transmitted or not. LEDs will be automatically on if the messages have arrived. LEDs are light sources that can convert an electrical energy into light. They are not difficult to be programmed and do not cost much, and they come in different colors. For this study, four different types of colors are used.

Buzzer: It is a device that is used to create a beep noise whenever a message is delivered to the Arduino Uno [1].

Power supply: Batteries are used to supply the Arduino with the power it needs to work. By using batteries, there is no need for electricity in the study. We did not use the liquid crystal display (LCD) screen in this study because of the lack of pins on the Arduino Uno necessary to connect the screen [1]. Therefore, the LCD screen was replaced by lighting LEDs to show the message sent by the patient.

3.3 Working principle

Firstly, the patient is required to wear a glove that will contain the first Arduino and the board. The mechanism of the gloves is that, according to the movement of the hand, a message will be sent. This device contains a specific sensor that will respond to the movement of the glove and transfer it into a type of signal; and these signals will then be translated into a message by using different types of sensors, one of which is the nRf sensor [16]. This nRf sensor is responsible for connecting the two implored Arduinos. In this study, we used two nRf sensors; the first one is used to send the signals coming from the accelerometer to the second one installed on the other Arduino. The accelerometer is programmed to respond to the four movements of the glove: left, right, up, and down. We also used four different colors of LEDs in order to recognize the different signals sent by the accelerometer. For example, we used the green light for left direction, the red for the right direction, the white for the upward direction, and the yellow for the downward direction. Thus if the accelerometer receives a signal for left movement, the green light turns on while the rest of the colors are off. Finally, the Arduino needs a power supply to work; in this regard, two batteries are used, each connected to the Arduino to prevent the need for electric wires.

3.4 Connections and problems

At first, we started the connection on the first Arduino Uno by connecting the accelerometer, which has six pins; however, we used four pins as needed (GND, VCC, x-axis and y-axis). We connected the GND to the ground then VCC to the +5V, the x-axis to analog input pin (A0), and the y-axis to analog input pin (A1). Afterwards, we connected the nRf sensor, which has seven pins [16]. The GND was connected to the ground and the VCC to 3.3V pin, and then, we connected the rest of pins to the Arduino by using the following pins: D7, D13, D12, D8, and D11 (Fig. 3.1).

After we finished the connections of the nRf sensor, we further connected the LEDs and the buzzer, and we used the breadboard to connect them altogether. On the breadboard, we placed the LEDs, which have two legs (anode and cathode). The cathode legs were connected to the negative side and the anode to the pins on the Arduino [1]. Since we had four LEDs, we used four pins from the Arduino (D3, D4, D5, and D6). Moreover, the buzzer was connected in the same way, one leg to the negative and the other to the pin (D2) on the Arduino. Finally, we connected the negative side of the breadboard to the ground on the Arduino Uno (Fig. 3.2).

Fig. 3.1 shows the finishing of the whole circuit. Once, it was finished, we ran the codes in the program in order to run both circuits simultaneously using the nRf sensor [16]. We also inspected the proper working condition of the whole system, and finally, we placed the first circuit on the wearable glove (Fig. 3.3).

From the circuits shown in Fig. 3.2, we faced some problems with the connections in Fig. 3.1. This issue was about the replacement of the LCD screen with some LEDs due to lack of some necessary pins in the Arduino Uno. The other problem that we faced was related to the nRF sensors [16]. There were some false statements in the code we uploaded to the Arduino, and after vigorous debugging, we determined the right code for the nRf

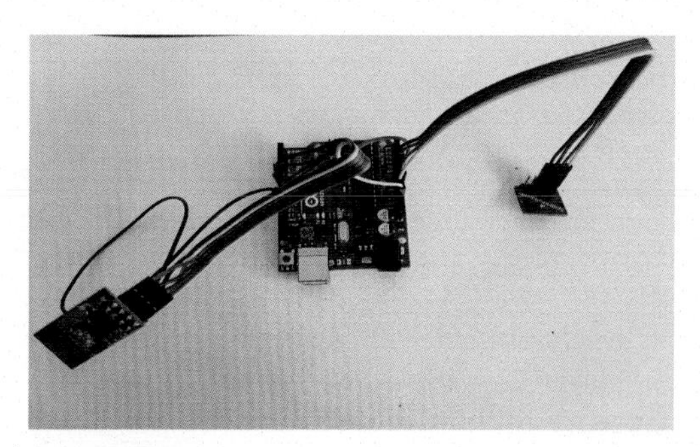

Fig. 3.1 First Arduino circuit.

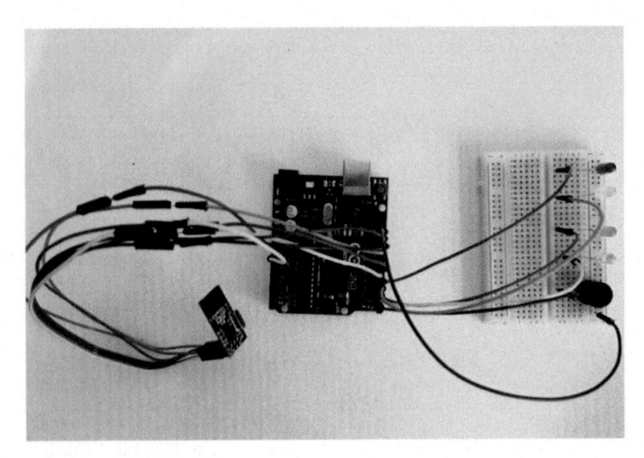

Fig. 3.2 Second Arduino circuit.

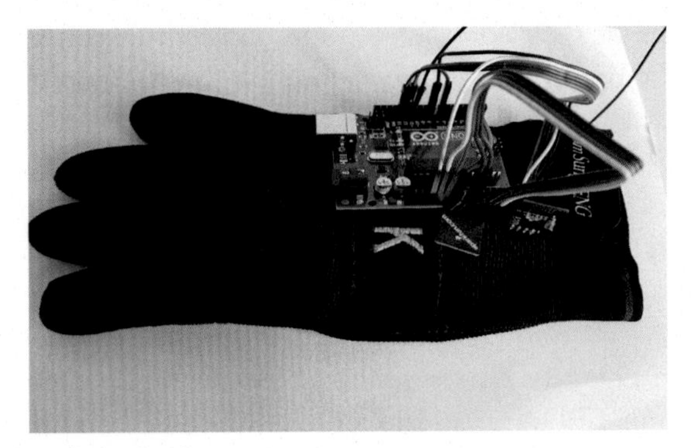

Fig. 3.3 Assistive glove circuit.

sensors; and we uploaded it again, and it started working properly. Before uploading the right code, the nRf sensors were not able to connect together so each of the nRf sensors was working separately.

Moreover, one more problem we faced was related to accelerometer sensor. This issue was related to its stability number which needed to be changed for proper performance. Figs. 3.4 and 3.5 depict the complete workflow of the first and second Arduino circuits.

3.5 Discussion

The advantages of this study can be its simplicity, efficiency, and easiness for both patients and caretakers. This is due to its reliability on a simple

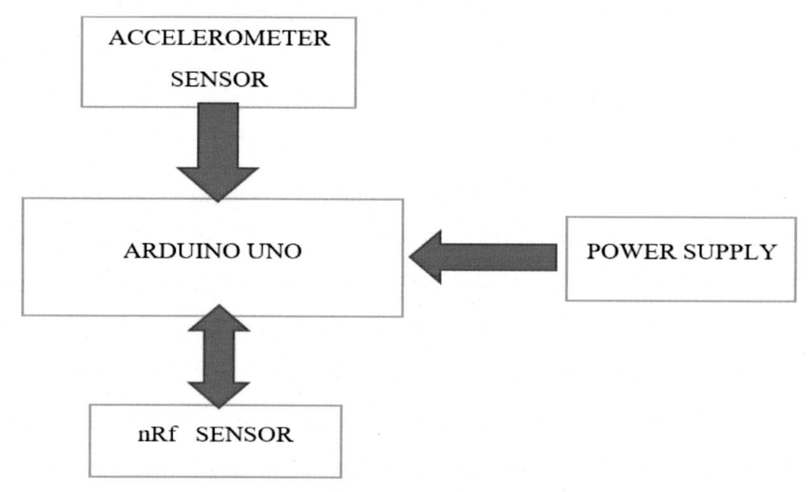

Fig. 3.4 The block diagram of the first Arduino circuit.

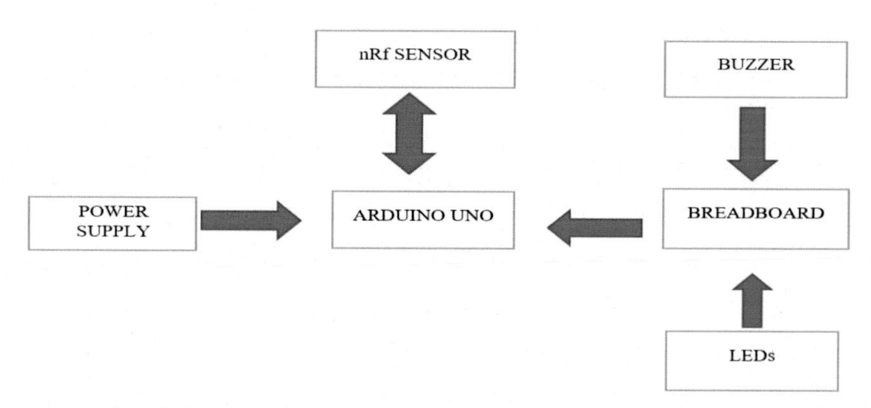

Fig. 3.5 The block diagram of the second Arduino circuit.

movement of the patient, which is leaning the hands, that results in sending a message directly to the caretaker.

Moreover, one of its advantages is the low cost. The proposed technology is less expensive to obtain because most of the materials we acquired are cheap and easy to acquire. In addition, with the usage of battery, it is unnecessary to use constant electricity output; hence it provides the patient with the free will to sit anywhere with no health hazard from an electricity source. Furthermore, it is a timesaver for both patients and caretakers because in using this device, a message will instantly be sent to the caretaker; and when the caretaker observes that message, he can see what the patient needs and provide him with requested needs instantly without delay.

The main disadvantage of this study is the limited functions of the device as it has limited messages to display on the LCD screen. We programmed the device on just four basic needs of the patient. Moreover, constantly charging the battery is another disadvantage encountered with novel technology.

Finally, this designed device, as earlier mentioned, is only good and applicable for the patients who have partial paralysis.

3.6 Conclusions

Paralysis is a disease that affects humankind and makes their life worrisome. By building such a device, we intend to make the lives of those affected by paralysis much easier, save unnecessary time wastage, and provide them with their essential needs. Our study consists of two Arduinos and different types of sensors besides some LEDs and a buzzer. Its final design is a glove that the patient wears, and according to the movement of the glove, one of the LEDs will turn on and be responsible for a specific message read by the caretaker.

The proposed device is designed to be easily used by the patient and less expensive because of the materials used to create it. In spite of being helpful, unfortunately, it has some disadvantages, such as its limited sent messages as it relies on the direction of the gloves and LEDs. However, this can improve.

3.7 Future aspects

This technology is made to be simple and easy to use. Some types of sensors were used to provide response to the movements of the hand (left, right, up, and down), so therefore we had to focus on giving only four messages that can be sent to the paralytic patient's caretaker. However, this can be improved. The first idea to be incorporated in the nearest future is to replace that accelerometer with another one that can recognize more than the four directions we proposed. This will make the device capable of sending more than four messages; for example, it could recognize the movement of the five fingers, and each one could have a different message. Also, in the glove device, we used LEDs to show what kind of messages arrived or what the patient needs; this leads to the idea of changing LEDs and replacing it with a screen, so it would be easier for the caretaker of the patient to read the message quickly and easily. Moreover, the proposed design is a wearable glove that can uncomfortable at times for the paralytic patient to wear. Hence this can be replaced by a smaller and more comfortable technology, such as bracelets, instead of the gloves. Also, the range of coverage of the nRf sensor is approximately 100 m. This can be improved by proposing another sensor with wider coverage. A GPS can also be added to the device in order to locate the patient's location whenever a message is delivered.

References

[1] https://store.arduino.cc/usa/arduino-uno-rev3. (Retrieved 15 December 2020).
[2] B.S. Armour, E.A. Courtney-Long, M.H. Fox, H. Fredine, A. Cahill, Prevalence and causes of paralysis—United States, 2013, Am. J. Public Health 106 (10) (2016) 1855–1857.
[3] M. Efeoglu, H. Akoglu, T. Akoglu, S.E. Eroglu, O.E. Onur, A. Denizbasi, Spinal trauma is never without sin: a tetraplegia patient presented without any symptoms, Turk. J. Emerg. Med. 14 (4) (2014) 188–192.
[4] G.M. Cooper, The development and causes of cancer, in: The Cell: A Molecular Approach, Sinauer Associates, Sunderland, MA, 2000.
[5] B.C. Marsh, S.L. Astill, A. Utley, R.M. Ichiyama, Movement rehabilitation after spinal cord injuries: emerging concepts and future directions, Brain Res. Bull. 84 (4–5) (2011) 327–336.
[6] T.M. O'Shea, Diagnosis, treatment, and prevention of cerebral palsy in near-term/term infants, Clin. Obstet. Gynecol. 51 (4) (2008) 816.
[7] N. Attal, G. Cruccu, R. Baron, et al., EFNS guidelines on the pharmacological treatment of neuropathic pain: 2010 revision, Eur. J. Neurol. 17 (9) (2010) 1113–e88,- https://doi.org/10.1111/j.1468-1331.2010.02999.x.
[8] S. Ramchandren, M. Jaiswal, E. Feldman, M. Shy, Effect of pain in pediatric inherited neuropathies, Neurology 82 (9) (2014) 793–797, https://doi.org/10.1212/wnl.0000000000000173.
[9] A. Girach, T.H. Julian, G. Varrassi, A. Paladini, A. Vadalouka, P. Zis, Quality of life in painful peripheral neuropathies: a systematic review, Pain Res. Manag. 2019 (2019).
[10] E. Olunu, R. Kimo, E.O. Onigbinde, M.A.U. Akpanobong, I.E. Enang, M. Osanakpo, A.O.J. Fakoya, Sleep paralysis, a medical condition with a diverse cultural interpretation, Int. J. Appl. Basic Med. Res. 8 (3) (2018) 137.
[11] S. Zhang, B.B. Lachance, B. Moiz, X. Jia, Optimizing stem cell therapy after ischemic brain injury, J. Stroke 22 (3) (2020) 286.
[12] C.V. Borlongan, Concise review: stem cell therapy for stroke patients: are we there yet? Stem Cells Transl. Med. 8 (2019) 983–988.
[13] A.E. Groot, J.D.M. Vermeij, W.F. Westendorp, P.J. Nederkoorn, D. van de Beek, J.M. Coutinho, Continuation or discontinuation of anticoagulation in the early phase after acute ischemic stroke, Stroke 49 (7) (2018) 1762–1765.
[14] L.M. DeAngelis, Brain tumors, N. Engl. J. Med. 344 (2) (2001) 114–123.
[15] L. Galluzzi, L. Zitvogel, G. Kroemer, Immunological mechanisms underneath the efficacy of cancer therapy, Cancer Immunol Res. 4 (2016) 895–902.
[16] https://developer.nordicsemi.com/nRF_Connect_SDK/doc/latest/nrf/include/bluetooth/mesh/sensor.html. (Retrieved 10 January 2021).
[17] https://www.fierceelectronics.com/sensors/what-accelerometer. (Retrieved 20 November 2020).

CHAPTER FOUR

Development of smart jacket for disc

Dilber Uzun Ozsahin[a,b,c], Abdulrahim S.A. Almoqayad[b], Abdullah Ghader[b], Hesham Alkahlout[b], John Bush Idoko[d], Basil Bartholomew Duwa[b], and Ilker Ozsahin[a,b,e]

[a]DESAM Research Institute, Near East University, Nicosia, Turkish Republic of Northern Cyprus, Turkey
[b]Department of Biomedical Engineering, Near East University, Nicosia, Turkish Republic of Northern Cyprus, Turkey
[c]Medical Diagnostic Imaging Department, College of Health Sciences, University of Sharjah, Sharjah, United Arab Emirates
[d]Applied Artificial Intelligence Research Centre, Department of Computer Engineering, Near East University, Nicosia, Turkish Republic of Northern Cyprus, Turkey
[e]Brain Health Imaging Institute, Department of Radiology, Weill Cornell Medicine, New York, NY, United States

4.1 Introduction

Low back pain (LBP) is recorded as one of the most predominant disorders with a high spread of about 70% among Americans. This leads to a low movement of nutrients in the body, which may subsequently cause a degenerative disc disease (DDD). Hence, this disease is recorded at a high rate among aging persons and people affected by accidents, bad posture, and/or cartilage endplate microvasculature. Common manual therapies have been applied hitherto before the evolution of technology. This method can also be exhausting, time consuming, and sometimes expensive. It requires someone's effort and attention to assist. In addition, individuals have complained of having an addiction to drugs prescribed to manage this disorder. Furthermore, many researchers and scientists have proposed a better approach to solve this medical condition using simple biomedical processes that are easy to use, cheap, and reliable. Smart Jacket is proposed to be efficient in carrying out these processes.

The lumbar, or lower back region, is an interconnected structure that connects bones, muscles, ligaments, and body nerves and gives flexibility and support to the body. However, treating lower back pain and related herniated disc disorders could be risky. Manual and medical methods devised have proven alternatively relieving but risky, which may involve cutting and adjusting the body parts. Common symptoms include achy lower back,

Modern Practical Healthcare Issues in Biomedical Instrumentation
https://doi.org/10.1016/B978-0-323-85413-9.00003-7

burning pains at the thighs and lower feet, muscle tightness, and prolonged pelvis and back pain. One of the basic remedies applied in treating early lower back pain is to avoid activities that may be strenuous. Mechanical processes involve warmth or ice bath as well as application of an electric pad or a heat wrap on the back to relax the muscles. The heat and ice methods are applied in these regions to subside the pain. Many studies have devised alternative ways to treat and manage the disc-related back pain using technology [1, 2].

Recently, research studies using modern advance massage therapies were devised; these techniques involve the use of an electromechanical approach in building novel devices that may subsequently replace manual massage therapy. Such methods include the use of vibrating handheld devices and vibrating sits or chairs. These methods have high records of managing and alleviating this pain. Furthermore, this research introduces the application of a smart technology and the potential of a textile-derived material in treating back pain or disc regenerative disease, using different installed devices within the prototype. Drugs such as aspirin, ibuprofen and naproxen are registered therapies that alleviate the pain. However, some people may be allergic to these drugs [3].

4.2 Preceding works

The smart textile industry is considered as one of the most advanced and widely accepted technological sectors now applied in medicine. The application of smart textiles (wearables) is now explored in tremendous ways to assist in solving medical problems. Attempts are made in the construction of smart jackets to solve different medical problems using smart clothes. This is explored in many fields, such as artificial intelligence, material science, and electronics and optical.

Lee et al. [4] developed a smart jacket specifically for fixing low back disorder in women aged 50s–60s. They worked on the integration of heating and lighting functions on their device. Lee and colleagues recorded excellent results using three main survey questions. They analyzed the device performance based on the convenience, practicality, and appearance. This device gave accurate results even after laundry. Hence, the device was developed to assist older women using a built-in vibration [4]. Similarly, Siddhartha co-authors [5] developed an incredible smart jacket for visually impaired persons to assist them in navigating freely. The developed smart jacket was composed of an embedded sensor, smartphone with four submodules, such as a power supply, sensor, communication, and microcontroller. Their result was sampled using an object placed 3 mm in front of the device. The jacket made buzzing sounds, warning the user of the danger ahead. The smart jacket consisted of an Arduino microcontroller system, a built-in power system, a buzzer, a sensor, a light-emitting diode (LED), and a Bluetooth module; and it had an accuracy of 98% [5]. Additionally, Rajendran et al. [6] designed a smart vest with a metal detector for personal protection used in industries. The design involves the use of rechargeable batteries, buzzers, resistors, transistor, and capacitors incorporated within a circuit board. They also incorporated a metal detector within the jacket that induces an alarming sound whenever a metal hazard is sensed. Questionnaires were completed by respondents after a survey regarding the satisfaction of the jacket design, simplicity, and light weight with an acceptance rate of >3.00/5.00. However, respondents suggested a better design and closer range sensor for the metal detection [6].

Randhawa et al. [7] developed a smart jacket to protect the body against bad posture. This smart jacket was designed using fabric sensors within the jacket to assist the body in detecting danger that may be detrimental. The device was developed to remind the user of the wrong state of the body; it incorporated different applications within to assist in getting a desired product. Such devices included pressure sensors, accelerometer, and stretch sensors placed on the right body location to respond to the changes in posture. The fabric sensors were sewn into the fabric using a conductive thread. The data was integrated on the finished prototype locating them at different parts of the jacket [7]. Bouwstra et al. [8] from the University of Technology, Eindhoven Netherlands, developed a smart jacket as a neonatal monitoring system in the intensive care for tiny newborn babies. It was designed to monitor the state of health of the newborn baby. Devices used in the development of the smart jacket included an ECG monitoring device, power supply, and sensors. The prototype device was considered accurately effective in carrying out the desired activity [8].

4.3 Block diagram of the proposed system

The novel prototype smart jacket developed for disc-related conditions is similar to the studies done by Siddhartha et al. The smart jacket is developed as an advanced massage therapy to manage the disc-related disorder. Fig. 4.1 depicts the block diagram based on the Siddhartha et al. (2018) design [5].

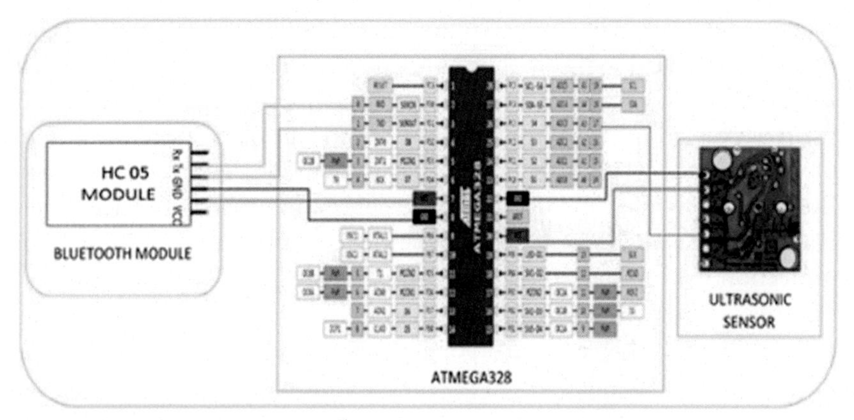

Fig. 4.1 Block diagram based on Siddhartha et al. (2016) design.

The architectural design was created using organized electromechanical processes combined in a fabric: (1) Microcontroller, (2) DC12V massage cushion vibrating system, (3) Bluetooth HC05, (4) DC Fan, (5) EMG, and (6) Sound sensor module. Details of which are demonstrated as follows:

1. Arduino: The ATMEGA 2560 microcontroller containing 54 connectors (outputs and inputs) was selected. It is a high performing RISC-based controller that contains 256 KB Flash memory, 8 KB RAM, and 8 KB of EEPROM. It contains 32 registers, 7 counters, and 16 analog to digital converters. In this study, the Arduino transfers data using sensors and power through a 5 V or 3.3 V outlet that can power up to 50 mA. The ATMEGA 2560 microcontroller is perceived to be cheap, tiny, and less power consumptive.
2. DC12V massage cushion vibrating system: the DC12V massage cushion vibrator uses a unipolar motor that reverses the electrodes without expecting flaws in reading. The massage device possesses special characteristics that are unique, such as speed, simplicity in use, and movement in nondisplaceable positions. These characteristics made it easier for this research. The DC12V massage cushion vibrator was incorporated into a fabric connected to sensors, buzzers, and a Bluetooth module for a proper data transfer [9].
3. Bluetooth HC05: Bluetooth module HC05 was used in this research study; this device allows the transfer of information to the smartphone. Bluetooth is regarded as a model necessary for the transfer of information from different devices connected to the jacket to the phone. The model Bluetooth HC05 was specifically picked for its high speed in

transferring data. Other special features include high level of sensitivity to about 80 Ms dB, wireless power transmitting signal +4 Ms dB, simple battery consumption, output signal 1.8–3.6 V, ability to control activities via ATtention commands integrated on the motherboard and connection ports for easy connection to the mail boards. The Bluetooth module operates in the form of serial communication. The Android program sends data to the Bluetooth module once a button is pressed and receives data from a device. This process assists in sending data to the Arduino panel via the transmit pin of the model for interpretation.

4. DC fan: The DC motor is installed with blades to act as a fan that controls the device. It operates by having constant current to reverse its movement. The reverse movement occurs when the opposite of the poles rotates in the opposite direction [10]. Relay terminals are connected to digital ports to make the fan spin clockwise; we made the first output value high and the second value low. For its counterclockwise functions, we reversed the previous process, either to make it stop or to make both sides low. The fan is programmed to rotate in the first direction for 3 seconds, and then stops for 3 seconds. Afterwards, it spins in the opposite direction for 3 seconds then stops for another 3 seconds, and so on until the current is disconnected.
5. Electromyography (EMG): EMG is a diagnostic procedure for assessing the safety of muscles and neurons controlled by them (motor neurons). The motor neurons transmit electrical signals that result in muscle contraction. EMG translates these signals into graphs, sounds, or numerical values for interpretation by a specialist. EMG uses small devices called electrodes to transmit or detect electrical signals. During the EMG process using a needle, the needle pole inserted directly into the muscle registers the electrical activities in this muscle. The neurotransmitter examination is another part of the EMG that uses electrodes connected to the skin to measure the speed of signals moving between two or more points. The EMG process interprets data derived from the body to the Bluetooth.
6. Sound sensor module: The audio sensor module is a simple microphone based on the LM393 power and the electro microphone. It can be used to detect any sound that exceeds the threshold, and the threshold can be adjusted from onboard voltage. Adjust the nautical voltage to adjust the sensitivity. There is high sound at the output unit when the sound is higher than the set threshold of high voltage. The outputs of this sensor can only be felt when it appears to exceed the threshold, but not in the amplitude and frequency.

7. RemoteXY program: With a mobile device that has either Android or iOS, it is possible to control a mobile device via a standard app in the Google Play store or AppStore. However, this often forces the user to use a program whose structure cannot be changed. RemoteXY is an application designed to transform this challenge into ease. With Arduino, an Android or iOS mobile device, and RemoteXY, it is possible remotely control desired peripherals in a wireless environment and set up fun and efficient applications easily by reading the RemoteXY program that allows users to design their own mobile device interface to enable wireless communications between mobile devices and cards, Arduino, and data control of sensors. Indicators can be placed on the screen that display the status of an Arduino task. After creating the interface in Visual Editor, the source code can be obtained from the Arduino program. The drawing source code that is downloaded will be finished directly in Arduino or complemented by a user's tasks. The Bluetooth module must be connected to the motherboard to establish wireless communications between peripherals.
8. Jumper wires were used in making our connections between items on the breadboard and the Arduino header pins. And there are different sizes of jumper wires (tall and short wires) to wire up all of the circuits.
9. Buzzers: The buzzer consists of an outside case and two pins, one for power and the other for ground. Inside the case, there is a piezo element that consists of a central ceramic disc and metal around it, so when force is applied to the buzzer, the ceramic disc will vibrate; and this vibration is what we hear. The vibration will either change to go faster or slower, which changes the resulting sound, so essentially, it is a speaker that can connect directly to the breadboard. We used it to alert the body of any changes in the sitting position that may subsequently lead to disc-related disorder.
10. Breadboard: There are three different types of prototyping boards (breadboard, per board, strip board). For each one, there is a special function. Breadboard is a construction base for prototyping of electronics, and there are three different sizes of the breadboard (full size, half size, and mini size). We used one full-sized breadboard and one half-sized breadboard in this research to make our connections. Because the Arduinos don't have enough spaces to make all the connections between wires in that order, we considered breadboards [11–13].
11. Medical back strap: This is a supportive belt for the back muscles, especially useful for those persons who deal with back pain, cartilage glands,

and lumbar spine pain that lasts for more than a week due to a number of reasons including weight gain, excess pressure on the back and lumbar spine, or occupation-related causes. This unit can reduce morbidity by 80% without the need for medical intervention.

Benefits of the medical back strap: It relieves the pain of sciatica caused by cartilage and pressure on the nerves of the lower limbs of the back and the muscles and strengthens the body and back to assist in walking properly. The vertebrae differentiate between the lower limbs of the back and the muscles by the air that is extended inside the belt, reduces numbness and muscle contractions, and gives a feeling of relaxation and comfort. This strap also relieves the burden of cartilage sliding between the vertebrae; it also prevents exposure to cartilage glaucoma for people who drive cars for extended periods. Relieves cartilage, cervical, lumbar, and lumbar punctures, relieves neck pain and protects ligaments from excessive stress.

The back belt may not dispense with the treatment of the condition, but it will be a wonderful addition to the system of treatment and will speed healing, acting as one of the most appropriate complementary treatments in the following cases: postsurgical procedures, as the patient may need a pillar to help in identifying and reducing movement to accelerate healing in the back and abdomen; slippage and spondylitis, to improve the ability to walk, relieve pain, and relieve damage to the nerves, muscles and ligaments of the surrounding area; osteoporosis, to relieve joint pressure and alleviate the pain associated with the disease; and spinal fractures, to help relieve the pressure generally on the spine [14].

The board driver motor: Integrated circuit L298 is a circuit used to drive two electric motors with a maximum value of 2 amps for each engine and is easy to control and carry volts up to 36 V. This integrated circuit consists of several entries. For motor entrances, the outputs are divided into 4 entries and divided by 2 inputs on each side, and they are connected by the electric motor. These outputs are numbered OUT1–OUT4. The first motor is connected to the OUT1, OUT2, and OUT4. The OUT4 is located toward the engine rotation, and the entrances are ENA and ENB. If the LOW is applied, the engine will never turn off and will stop. If the HIGH mode is set, the control goes through the inputs numbered IN1–IN4. In general, the pins jumper is small and between these portlets with the 5 V to activate it all times. Concerning control entries, there are IN1–IN4. IN1 controls the input, and OUT1 and IN2 controls OUT2. Thus by applying 1 logical 5 V on the IN1 input, for example, the current will be passed through OUT1, turning off the current through OUT1. Since the L298N is a fully integrated dual-channel circuit, you can control two engines at the same time with separate control legs for each engine. The logical voltage required for the circuit is 5 V, but the engine feeding voltage can reach up to 45 V. The output current peak is also considered for each 2 A channel. In general, the L298N is available as a unit containing all the necessary components and connections to control two motors. The L298N circuit module consists of a two-pole connection to two diodes for connecting two engines and has six memo electrodes: two for input inputs, four for input control (two for each drive).

The running circuit controlling the medical back strap is composed of:

1. energy providing device (power bank)
2. microcontroller board (Arduino MEGA)
3. fan
4. Bluetooth device
5. breadboard and wires for connecting the circuit components

Power bank: The main function of the power bank is providing the whole circuit with the energy needed to run it and complete its function. The type of power bank must be at least with a capacity of 10,000 mA to ensure a long duration for working time. It should not be less than 9 V to be capable of running the circuit depending on the connected devices.

Two types of Arduino found:

1. Arduino MEGA
2. Arduino NANO

The components of the Arduino MEGA:

1. 54 digital inputs/outputs
2. 15 are pulse with modulation
3. 16 analog inputs
4. 256 KB of flash memory
5. 8 KB ram

The Arduino MEGA has more functionality and advantages than the NANO type, so it was used instead of the NANO.

Basics of resistors: Resistors are electronic components with constant static electrical resistance. The resistors restrict the passage of electrons (current) through the circuits.

They are passive components, meaning they only consume energy and cannot generate it.

Often resistors are added to circuits along with active components such as op-amps, microcontrollers, and other integrated circuits. The most common uses of resistors are limiting current and dividing voltage.

Resistors units: Electric resistance is measured by ohm and is symbolized by omega (Ω). The ohm defines electric resistance between two points if a voltage difference of 1 V is applied between them, resulting in a 1 amp. Like other SI units, larger and smaller values of resistance are specified using prefixes, such as kilo, mega, or gigabyte to denote large values, or mille and micros to denote small values. It is common to see resistors in kilo-ohms (kΩ) or mega-ohms (MΩ) while resistors are less than a milli-ohm (mΩ). Fig. 4.2 is a caption of the explored resistor.

Resistors have many shapes and sizes; they can be surface mounts or mounted through holes.

Plug and install: Resistors come in two types (in terms of installation): superficial or installed through holes. These types are often abbreviated to SMD (surface mount device) and PTH (plated-through-hole), respectively. The resistors installed through the holes have two flexible wires that can be installed on the breadboard or can be manually welded into the prototyping board or printed circuit boards (PCBs). These resistors are usually more useful with test boards and prototyping or in any other case where you do not want to install tiny resistors (0.6 mm) of surface resistors. These resistors often need to be pruned and then fixed to a large area compared to their surface resistors. The most common types of resistors installed through the holes are the axial package; the size of these packages is proportional to their power rate. For example, the resistor has a half-watt rate of about 9.2 mm while the resistor has a quarter-watt rate of about 6.3 mm [15].

Configuration of resistors: Resistors can be made from many materials, and the most common modern resistors are manufactured from carbon, metal, or metal-oxide film. In these resistors, a thin layer of conductive material (but having high resistance) is wrapped around a dielectric material and covered with the same material. Most of the standard resistors installed through the holes consist of a layer of carbon or a layer of metal.

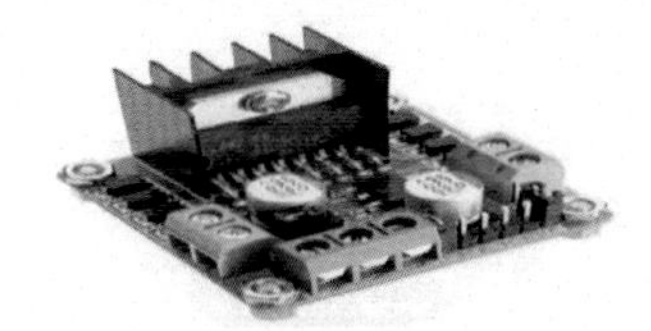

Fig. 4.2 Resistor.

Explanation of symbols and signs of resistors: Although their value is not explicitly written, most resistors display their value of their electrical resistance. PTHs are distinguished by a color coding system, and surface resistors (SMDs) have their own coding system.

Interpretation of color bands: The first two rings indicate the value of the first two digits of the value of the resistance; the third loop gives the value of the decimal multiplier of the resistance (i.e., the first value is multiplied by the value of the resistor). Two rows of 10 rose to the strength of the value of the third episode. The last episode shows tolerance, the amount of increase or decrease of the actual value compared to its nominal value. There is no resistance made perfectly, and different manufacturing processes result in an increase or decrease in the value of resistance. For example, a resistance of 1 kΩ and a 5% permittivity has an actual value ranging from 0.95 kHz to 1.05 kΩ.

How is the distinction between the rings (first and last)? The last ring of the emeralds is placed away from the rest of the rings in a distinctive manner, and usually the color is silver or gold.

Symbols of surface resistors: Surface resistors (such as the packages already mentioned: 0603 or 0805) have a special system for displaying their value. There are some common ways to write down resistors that will be shown on the surface of these resistors. There are usually three or four symbols (letters or numbers) printed on the surface of the resistor. If all the symbols you see are numbers, it means the resistor is an E24 series resistor. This system is somewhat similar to the color ring system used in the mounted resistors through the holes [16].

Advantages of using a smart jacket for a disc-related disorder/massage:

A. Relaxation

The use of an electrotherapy device usually comes at the end of the massage session and is used for more relaxation and stimulating blood flow before the massage finishes. The vibration movement relaxes the muscles and creates a feeling of relaxation on various parts of the body.

B. Pain relief

Vibration massage can be used as a means of relieving pain. It does not include medication. Vibrating massage helps relieve painful conditions by creating numbness in the affected area. Vibration can also help to relax the muscles surrounding the pain site. This reduces pain if you cannot apply vibration directly to the affected area.

C. Recovery

Vibrating massage is used with an electric massage device to provide the receiver with a sense of recovery and relaxation. Massage can be performed with a mild vibration before any sporting event to stimulate blood flow and leave an athlete feeling a sense of energy.

Risk of using an electric massage device

1. Treatment with an electrocardiogram may be dangerous if the vibration intensity is very high; this may cause lower back injuries and severe back pain.
2. Impromptu sound of the buzzer and sensor may cause shock to the user.
3. Electrical faults on the jacket may affect the user.

EMG test: This test is performed by inserting electrodes into the intramuscular muscles to record the electrical potential of the muscle fibers. In most cases, a concentric needle is inserted so that its center is isolated from its circumference while the measured potential is between the center and the periphery. The electrical activity of the muscles is documented on the EMG screen and can be heard through the loudspeaker. The combination of what the eye sees and what the ear hears is important and helps those who are being tested to assess the quality of the muscle's electrical activity. The electrical properties of muscle fibers, in professional language, are called shrapnel.

The electrical activity of the muscle is measured in three cases: In the muscles working properly, there is no inhibition, the performance of the fibers is determined between its active and relapse states. This occurs with fibrillation from fibers with defibrillation, or acute potential. When the muscle is slightly activated, it appears on the screen when exposed to neuromuscular or neurological diseases. The size of the motor units increases and appears on the screen. However, when exposed to muscular diseases, these motor units become smaller.

Frequent nervous stimulation: This test is done by evaluating diseases affecting the nerves and muscles, such as muscular dystrophy (myasthenia gravis). Similar to the way this test is performed, it appears that individual stimuli stimulate the nerves several times. However, when exposed to musculoskeletal-related diseases, stimulation is performed at a frequency of 2–3, a gradual decrease in the magnitude of the recorded reaction occurs [17].

The circle of massage: There are 13 ports, Arduino sensor shiled V4 and V5, ground (GND) and voltage common collector (VCC) in the board.

The server is responsible for the board feeding. For the connection, at first, we linked the negative axis with the negative in the board and the positive with the positive for the first and second stages. Secondly, we connected the VCC with the positive battery of 12 V or 9 V, and no problem was encountered; then we connected with Bluetooth and Arduino with a negative battery. We further connected V5 with Arduino and Bluetooth to establish the communication between them. Thirdly, there are four entries for the new devices; the function is to control the speed of the Taurus, and we connected them with the Arduino IN1 with the 50, the IN2 with the dial 51, the IN3 with the 52, and the IN4 with the 53. Fourthly, there is a pint of the pwm function that controls the rotation of the Taurus. The first entrance is the end link connected with the entrance of the Arduino 8 pwm and also connected ENA with the entrance of the Arduino 9 pwm. The fifth is the Bluetooth, which has four inputs: the TX input; the RX input; the VCC input; and the RX input with the input to the Arduino 11 pwm. Sixth, Brukram worked through the RemoteXY interface design program to control the massage and fans and define each input for the joystick. Figs. 4.3 and 4.4 depict the TX input and RX input connectivity respectively.

Fig. 4.3 TX input.

Fig. 4.4 RX input.

The circle of EMG: The muscle sensor measures the electrical activity of a muscle and produces an analog output signal that can easily be read by a microcontroller. As a word of caution, this product needs a positive and negative reference voltage; two power supplies are required. The sensor has a maximum operating voltage of 18 V. However, based on this research, it is recommended not to use anything higher than 9 V to minimize the risk of electric shock. Electrolysis of muscles or muscle layout can reveal the results of medical planning of muscle dysfunction, muscle dysfunction, or problems in transmitting nerve signals to muscles. Benefits of electrolysis of muscles include restoration of strength and function, prevention of muscle atrophy, and reduction of muscle spasms. A doctor may ask patients if they have had an electrical muscle screening test if they show signs or symptoms that may indicate nerve or muscle disorders. These symptoms may include numbness, muscle weakness, muscle pain or cramping, and pain in the limbs. The results of electrical muscle planning are often necessary as they help diagnose or exclude a number of conditions such as muscle disorders, muscular dystrophy, muscle inflammation, and diseases that affect nerve contact points in muscles, such as myasthenia gravis.

Connections: We connected the power supply to 9 V batteries and connected the positive terminal of the first 9 V battery to the +Vs pin of the sensor. We also connected the negative terminal of the second 9 V battery to the Vs pin of sensor and connected the negative terminal of the first 9 V battery to the positive terminal of the second 9 V battery. Then, we connected to the GND pin of sensor, connected the SIG pin of sensor to an analog pin on the Arduino (A0), connected the GND pin of the sensor to a GND pin on the Arduino, and connected the SIG pin of the buzzer to a digital pin on the Arduino (Digital PWM 3). We also connected the VCC/GND pin of the buzzer to +5 V/GND pin on the Arduino. The yellow wire (light gray wire in print version) was pasted at the upper vertebrae of the spine first, and the red wire (gray wire in print version) was attached to the left side of the top of the back (Fig. 4.5). The green wire (dark gray wire in print version) was attached to the right side of the top of the back [18].

4.4 Result and discussion

Table 4.1 compared different smart jackets developed to manage different diseases. These devices were compared based on criteria, such as the type of the disease/disorder, simplicity, effectiveness, environmental friendliness, convenience, acceptability, appearance, and cost.

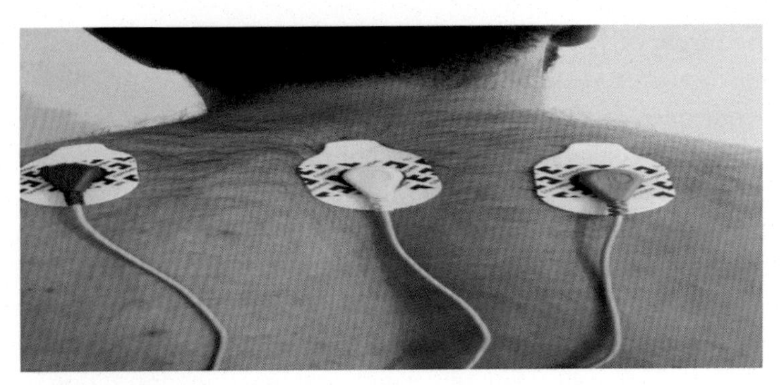

Fig. 4.5 EMG connection.

Our designed smart jacket device proved outstanding compared to other devices. The device is exceptionally effective and is recorded to be less expensive than the compared devices. Other factors that make our device better in comparison is the fact that it has no replica or exact developed device.

4.5 Conclusion

In this study, an ergonomic and convenient smart jacket device was developed for individuals suffering from lower back disc-related pains for massaging purposes. This device was developed to overcome the limitation of other devices built hitherto. It integrates alarming and vibration functions at the same time. Also, an efficient outlook with affordability makes it exceptionally great compared to other devices. The design of this novel smart device amounts to critically understanding the electromechanical organization of the device and the subsequent challenges involved in its application. Getting individuals to try wearing this prototype device was challenging, so we tried using it first. We propose building a better, more affordable smart jacket disc device that is miniaturized into unnoticeable fabric clothing and also consider making it look modern. A heart rate monitor could be implemented. It would allow for a measure or display of the heart rate in real time or recording the heart rate for later study. It would be largely used to gather heart rate data while performing various types of physical exercise.

Table 4.1 Result comparison of different smart jacket devices for different medical functions.

Research title	Function	Simplicity	Effectiveness	Environmental friendliness	Convenience	Acceptability	Appearance	Average cost (USD)	Reference
Rating (/5)		3.9	4.5	3.0	3.0	4.9	4.0	350 >	
Electronic smart jacket for the blind	Designed to assist the blind	4.0	4.9	4.0	4.0	4.9	4.0	500 >	[5]
Smart jacket for disc-related disorder	Degenerative disc disease	4.9	4.0	2.0	4.0	3.0	4.0	< 200	Propose study
Smart vest with metal detector	Detecting metals in construction sites	4.9	4.0	2.0	4.0	3.0	4.0	500 >	[19]
Smart jacket for posture detection	Jacket to correct body posture	4.0	4.0	3.0	4.0	4.0	4.0	350 >	[7]
Smart jacket for neonatal monitoring	Baby monitoring device	4.0	4.0	3.0	3.0	3.0	3.0	300 >	[8]
Wearable lymph massaging modules: Proof of concept using origami-inspired soft fabric pneumatic actuators	Massaging device	4.0	3.0	3.0	3.0	3.5	3.5	< 500	[20]

References

[1] Global Burden of Disease Study 2017 Collaborators, Global, regional and national incidence, prevalence, and years lived with disability for 354 diseases and injuries for 195 countries and territories, 1990_2017: a systematic analysis for the Global Burden of Disease Study 2017, Lancet 392 (10159) (2018) 1789–1858.
[2] S. Dagenais, J. Caro, S. Haldeman, A systematic review of low back pain cost of illness studies in the United States and internationally, Spine J. 8 (1) (2008) 8–20.
[3] P. Weerapong, P.A. Hume, G.S. Kolt, The mechanisms of massage and effects on performance, muscle recovery and injury prevention, Sports Med. 35 (3) (2005) 235–256.
[4] J.-R. Lee, K.-j. Paek, G.-Y. Kim, Development and evaluation of smart jacket for women aged fifties and sixties, Fash. Text. Res. J. (2011), https://doi.org/10.5805/KSCI.2011.13.6.926.
[5] B. Siddhartha, A. Chavan, B. Uma, An electronic smart jacket for the navigation of visually impaired society, Mater. Today Proc. 5 (4) (2018) 10665–10669.
[6] S.D. Rajendran, S.N. Wahab, S.P. Yeap, Design of a smart safety vest incorporated with metal detector kits for enhanced personal protection, Saf. Health Work 11 (2020) (2020) 537–542.
[7] P. Randhawa, V. Shanthagiri, R. Mour, Design and development of smart-jacket for posture detection, in: International Conference on Smart Computing and Electronic Enterprise (ICSCEE2018) ©2018, IEEE, 2018.
[8] S. Bouwstra, W. Chen, L. Feijs, S.B. Oetomo, Smart Jacket Design for Neonatal Monitoring with Wearable Sensors, 2009. December 14, 2020 at 15:07:45 UTC from IEEE Xplore.
[9] M. Abdulhamid, K. Njoroge, Irrigation system based on Arduino uno microcontroller, Poljopr. teh. 45 (2) (2020) 67–78.
[10] Disc surgery and pain, Back Lett. 35 (9) (2020) 97–106.
[11] A.A. Khan, Application of electromyography in rehabilitation devices, Ergon. Int. J. 4 (4) (2020).
[12] L. Spruce, Surgical jacket and bouffant use and surgical site infection risk, JAMA Surg. 155 (10) (2020) 996.
[13] D. Norman Craig, Electrolytic resistors for direct-current applications in measuring temperatures, J. Res. Natl. Bur. Stand. 21 (1938). August 1938. Research paper RP 1126.
[14] B. Ghosh, Short circuit, Cult. Crit. 108 (1) (2020) 200–208.
[15] https://learn.sparkfun.com/tutorials/resistors/all.
[16] B. Crooker, Degenerative disc disorder, JAMA 298 (10) (2007) 1136.
[17] Cardiac Electrophysiology Clinics, Card. Electrophysiol. Clin. 12 (2) (2020) xi.
[18] P. Hodges, Editorial: Consensus for Experimental Design in Electromyography (CEDE) project, J. Electromyogr. Kinesiol. 50 (2020) 102343.
[19] S. Kodam, N. Bharathgoud, B. Ramachandran, A review on smart wearable devices for soldier safety during battlefield using WSN technology, Mater. Today Proc. 33 (2020) 4578–4585.
[20] H.J. Yoo, W. Kim, S.-Y. Lee, J. Choi, Y.J. Kim, D.S. Koo, Y. Nam, K.-J. Cho, Wearable lymphedema massaging modules: proof of concept using origami inspired soft fabric pneumatic actuators, in: 2019 IEEE 16th International Conference on Rehabilitation Robotics (ICORR) Toronto, Canada, June 24–28, 2019, 2019.

CHAPTER FIVE

Evaluation of a finite element laminectomy

Dilber Uzun Ozsahin[a,b,c], John Bush Idoko[d], Basel Almagharby[b], Mohammed Bin Merdhah[b], and Ilker Ozsahin[a,b,e]

[a]DESAM Research Institute, Near East University, Nicosia, Turkish Republic of Northern Cyprus, Turkey
[b]Department of Biomedical Engineering, Near East University, Nicosia, Turkish Republic of Northern Cyprus, Turkey
[c]Medical Diagnostic Imaging Department, College of Health Sciences, University of Sharjah, Sharjah, United Arab Emirates
[d]Applied Artificial Intelligence Research Centre, Department of Computer Engineering, Near East University, Nicosia, Turkish Republic of Northern Cyprus, Turkey
[e]Brain Health Imaging Institute, Department of Radiology, Weill Cornell Medicine, New York, NY, United States

5.1 Introduction

Spinal disease known as dorsopathy, manifesting as weakness in the spinal cord, is a condition concerning damage of the spine. This includes different sicknesses of the spine such as excessive outward curvature of the spine.

Persons between the age of 50 and 60 and older, in which their spinal cord narrows, could suffer from back pain and several problems. The spine is a series of connected bones and shock absorbing disks; it protects and saves the spinal cord and the most important part of the central nervous system that connects the brain to the body. For most people, the narrowing results from changes either after surgery or before in daily life because of painful joint or ligaments swelling. The open spaces (gaps) between the backbones may start to shrink. The narrowing could nip the spinal cord or the nerves, leading to pain, numbness, or tingling mostly in the legs, arms, or middle part of the body, as shown in Fig. 5.1. There is no treatment for it unfortunately, but there is a great number of nonsurgical treatments and sports to keep the pain under control. Most people with spinal stenosis live a normal life but experience pain [1, 2].

The benefits of titanium metals are monumental, giving us free space to make different designs in many applications, not only in the medical field but also in automotive and electrical and mechanical designs. One of the most important benefits is the resistance to corrosion is high, and additionally,

Modern Practical Healthcare Issues in Biomedical Instrumentation
https://doi.org/10.1016/B978-0-323-85413-9.00001-3

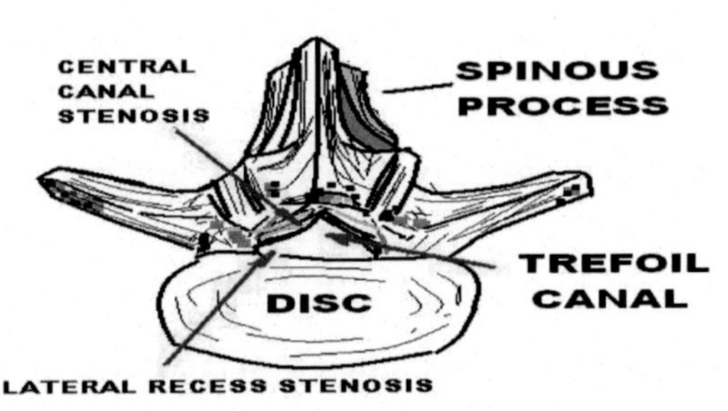

Fig. 5.1 Lumbar vertebra showing central stenosis and lateral recess stenosis.

these metals are biocompatible; and the human body, especially the bones, has the ability to accept this new guest. It is stable when inserted into the human body [3, 4].

It is known that the titanium and titanium alloy are very important metals in the medical field, whether as an instrument or as an implant in the human body because of its great characteristics. With these characteristics, engineers are able to design many types and produce them according to the situations and requirement of the patient [5, 6].

5.2 Causes

The cause of spinal diseases is basically joint swelling or joints disease; an event from the collapse of padded material in the gap between the vertebrae; and the expansion of bones causing disk changes, a concentration of the ligaments and bone nips. This can load pressure on the spinal cord and nerves. Other causes might be herniated disks. A person suffers a hernia if the cushions are damaged; material could leak out and compress on the spinal cord nerves, causing pain and some other risks to the patient. A sudden accident may crack a portion of the spine, leading to spine diseases. If cancer is detected on the spinal cord, the patient may get narrowing disease. Here, the bone grows in a way that is different from what is usually expected; growing large and becoming easily broken, the bone becomes week. The result is a spinal narrowing [7].

5.3 Diagnosis and tests

In most cases, the patient has to undergo one of these methods or devices for diagnosis.

X-rays: This type of device shows the structure of the backbones, and the doctor determines the patient's state after the X-examinations [7].

Magnetic resonance imaging (MRI): Using strong magnets and radio frequency leads to the creation of a three-dimensional (3D) image of the tissue to show the backbones and to show how much damage occurs in the bones. [7]

Computerized tomography (CT scan): A type of X-ray device that is used to create a three-dimensional image of the backbone by using X-ray photons that work similarly to the conventional form but in 3D form [7].

5.4 Treatments

The doctor may start off with nonsurgical treatments. In all situations, doctors try to do a nonsurgical treatment first. These might include medicine such as pain-relieving medicine, muscle relaxants, etc. [8].

5.4.1 Drug injections

The doctor injects a dose of medicine into a patient's back or neck to go through the spinal cord, which lessens swelling. However, because of the side effects, doctors use high quality pain-reducing drugs to stop the pain from advancing for some time.

5.4.2 Exercise

By doing some sports and exercises, the bones could heal faster, and they can improve the flexibility of the backbones leading to minimization of the stenosis for a long period.

5.4.3 Helping devices

Metallic supports may be utilized to help patients move well quietly, but this is temporary.

Some patients have extreme situations, including difficulties in walking as well as urine storage. Surgeons often suggest surgery for these kinds of patients. Surgeries such as a laminectomy create a gap in between the vertebrae.

5.4.4 Laminectomy surgery

This type of surgery is common, and most doctors find it as the best solution because it fixes the stenosis swiftly. The main purpose of this surgery is to target the nerve roots and to minimize the pressure since such pressure leads to leg pain and some other risks. It is also used to cure other situations, such as spinal cord injuries, injured disks, and tumors. In most situations, minimizing the pressure on the nerves can help the patient to feel better and to live a normal life. Laminectomy is a process of removing a bone portion of the backbones or tissue that is narrow and that squeezes the nerves. This surgery is done by surgically cutting the back of the spinal cord to deal with the vertebrae, and sometimes spinal fusion surgery may be done at the same time. Some surgeons prefer this step to help make the spine stronger with minimized pressure. Spinal fusion is an important surgery that usually does not take more than 2–3 hours. There are different methods of spinal fusion.

Sometimes a surgeon uses a bone either from a patient's body or the bone bank to create a bridge between spinal backbones. This will let the bone grow in a correct and useful form.

In most situations, an added fusion surgery is made by using metallic implants and metallic supports for the backbones, helping to align the bones together in the same level until new bone grows in that gap.

Ways of doing things change depending on the type of bones or metal implants that are used. The methods used always depend on some criteria such as the patient's age and health, how many backbones are under pressure, the extreme harshness of nerves under pressure, and the connected signs of sickness.

Spinal fusion is a surgery that leads the patient to heal in a very quick period and places them in a very good condition in a short period [8].

There is more than one stage in a laminectomy surgery. These include unilateral laminotomy, bilateral laminotomy, facet-sparing laminectomy, and laminectomy with complete facetectomy, as shown in Fig. 5.2.

The similarities among in vitro biomechanical research has not been reported yet. Eight lumbar spine medical samples were researched. Each medical sample was examined unharmed in one piece after two decompression procedures. All back parts were preserved in Stage A (Unharmed and in one piece). In Stage B (Bilateral), the lower of L4 lamina and upper of L5 lamina were removed, but the L4–L5 bone-to-bone connecting band was preserved. An opening was made on both sides. In Stage C, the laminectomy lower L4 and upper L5 were removed. Ligament yellow and ligament of L4–L5 were also removed. A machine is used to increase the moment up to 8400 N mm in elasticity. Spinal displacement when the compressive level L4–L5 is removed was measured by extensometer.

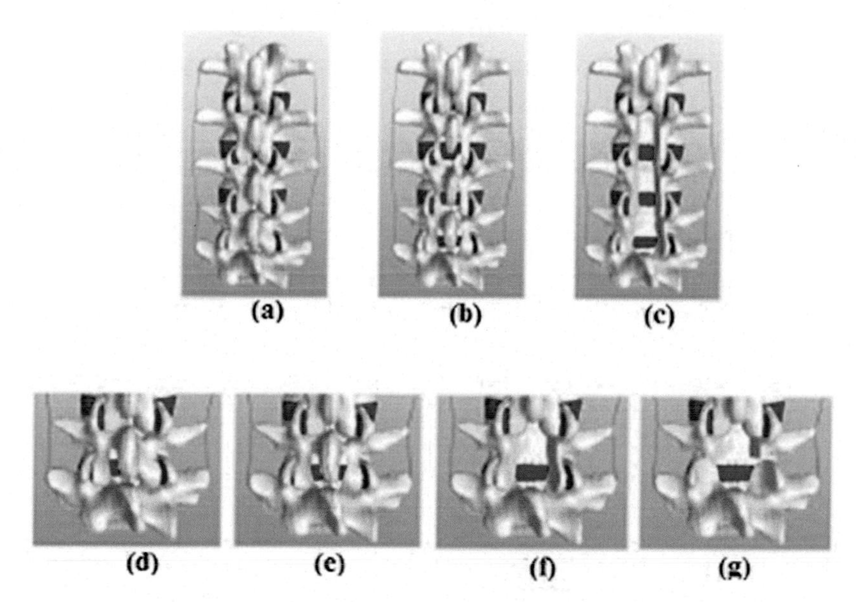

Fig. 5.2 Posterior view of types of laminectomy surgery: (A) Intact spine. (B) Bilateral laminotomy at L2–L5. (C) Fact-sparing laminectomy at L2–L5. (D) Unilateral laminotomy at L4–L5. (E) Bilateral laminotomy at L4–L5. (F) Fact-sparing laminectomy at L4–L5. (G) Laminectomy with facetectomy at L4–L5.

What to expect after surgery: Sometimes if the patient is younger, the possibility to get well sooner is high because it is expected that younger patients can do the recommended exercises and can move in a correct way, avoiding any wrong position of the spine. It likely takes 1–2 months for them to get back to the normal life [8].

Laminectomy, laminotomy, and spinal fusion are surgeries used to overcome the spinal diseases that cause some difficulties in walking and moving. If the patient chooses to use nonsurgical treatments, the pain will persist; some researchers describe the nonsurgical treatments as "time wasting." Also some patients cannot control their urine storage or bowel; thus, these kinds of diseases will put the patient in many difficulties, and if patients are young, they will feel older and will face some limits in moving because of the pain caused by the stenosis and spine narrowing. The surgeons always apply spinal fusion surgery at the same time because it will cause a short time period for the patient to heal, and it gives the patient the possibility to get back to normal life and to move freely [8].

How will it work: These types of surgeries sometimes may be recommended by the surgeons if the nonsurgical treatments will not work for the patient, if the doctor checks the patient and notices that the surgery will be helpful and can help the patient to overcome this problem, and if the doctor sees surgery as a patient's means of overcoming their walking and moving difficulties. For patients with extreme pain in their legs and nerves whose doctors advise them to do the surgery, this may help them a great deal. Also, if a patient is fat or otherwise considered obese, a doctor may provide a plan for weight loss to help ensure a successful surgery and postop healing period [8].

Risks: All operations on the spinal cord tend to be risky. These difficulties are more dangerous in older patients [8]. Some difficulties include cases from drugs that cause numbness or unconsciousness, large infection during the surgery, skin inflammation, coagulate in blood, unstable in movement, damage or harm in nerves, problems in urine, sac pain that takes a long time to heal, which is rare sometimes, and the possibility of death, though this is rare. Patients who smoke or who have a disease where their blood sugar swings may fall victim to these factors [8].

What do you need to think: Many doctors advise people with spinal diseases to first try nonsurgical treatments before choosing surgery. Surgery for the lumbar spine is most preferred to reduce pain and weakness within the legs. Surgery may not be more accurate to reduce the back pain.

5.5 Medical sample production

Eight lumbar spines from L1-S1 were used in this research. The para spinal muscles of each medical sample are cut out, all the ligament parts, including the supraspinal ligaments, were carefully examined. A bilateral laminotomy after the surgery, at different levels of cut out, and the compressive surgery are demonstrated in Figs. 5.3 and 5.4. All posterior parts were preserved unharmed and in one piece. In bilateral laminotomy, holes are made on both sides.

5.5.1 Biomechanical examination

The medical samples are chosen for biomechanical examination by MTS (Bionix 858, MTS Corp., Minnesota, United States). A uniquely designed fixture used to increase the moment up to 8400 N mm created over the pivotal movement of the MTS mechanical pushing-pulling device was applied to each medical sample to accomplish the elasticity movements. During the examination, intervertebral displacement at compression levels from L4–L5 was recorded by the MTS extensometer. During examination, intervertebral displacement data were concurrently recorded by the MTS TestarII

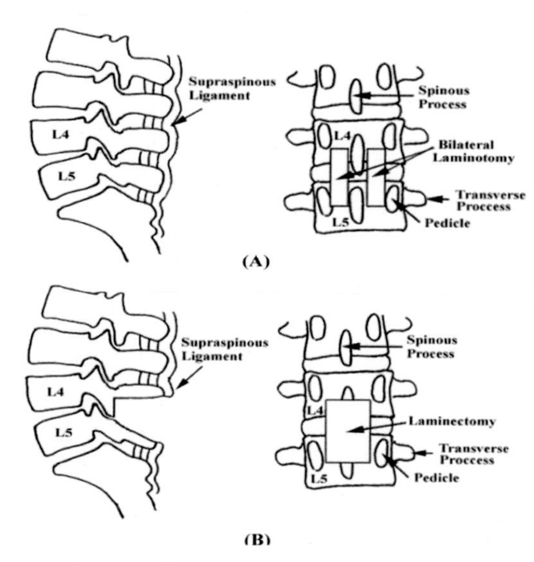

Fig. 5.3 Side view and view of lumbar spine after (A) laminectomy and (B) bilateral laminotomy.

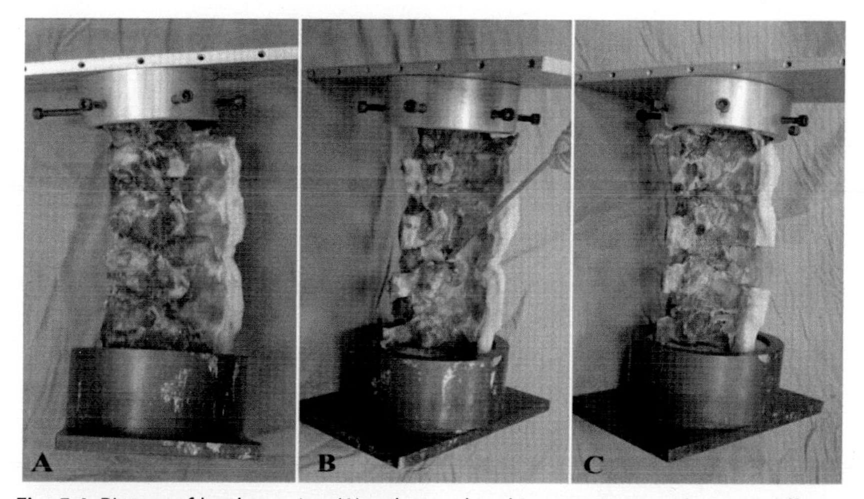

Fig. 5.4 Picture of lumbar spine (A) unharmed and in one piece, and at two different levels of taking out the compression following (B) bilateral laminotomy and (C) laminectomy.

software. Six intervertebral displacement measures of L4–L5 lumbar part were completed in each medical sample. Unharmed and in one piece during flexion, unharmed and in one piece during extension, bilateral laminotomy during flexion, bilateral laminotomy during extension, laminectomy during flexion, and laminectomy during extension were all examined.

For medical samples with bilateral laminotomy, we did the measurements unharmed and in one piece; the bilateral laminotomy was performed by opening a hole on both sides of the lamina. The lower of L4 lamina, upper of L5 lamina and the L4–L5 ligament yellow were removed. The ligament and supraspinal ligament of the L4–L5 were removed. The group of the lumbar spine unharmed and in one piece form, after bilateral laminotomy surgery and after laminectomy surgery, was rated using information from comparing results for intervertebral displacement between L4–L5, as shown in Fig. 5.5. The results depicted that during extension movement, intervertebral displacement between the medical sample in unharmed and in one piece form and at two different levels of cut out, the compression does not record much difference ($P > .05$). During flexion movement, vertebral displacement of the laminectomy medical samples at cut out, the compression level L4–L5 is by numbers more than in unharmed and in one piece or bilateral laminotomy medical samples ($P = .0000963$ and

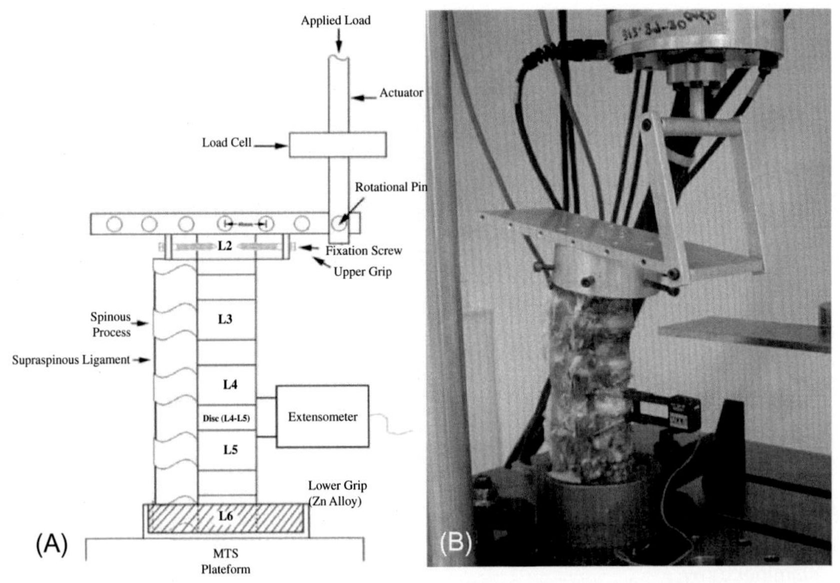

Fig. 5.5 (A) Experimental work for measuring intervertebral displacement and (B) elasticity movement of unharmed and in one piece lumbar spine. By using a uniquely designed shape, an 8400 N mm constant moment created over the axial movement of the MTS mechanical pushing-pulling device was applied to the medical spine sample to accomplish or gain with effort elastic movements. Intervertebral displacement is at L4–L5.

$P = .000418$, match up each pair of items in order). No difference was found between unharmed and in one piece and bilateral laminotomy ($P > .05$).

5.5.2 Aim and design goals

The bones were placed in a stable and fixed position; the handle was made strong to hold the loads of the body. The load and weight will be different by the size, mass and age of the patient, and situation of the bones. The tool must also be small and suitable for the patient, not to affect any muscle, bones, and soft tissue. If it is too large, it may cause harm to the tissue, leading to removal of the bone or inflammation. The surgeon must be precise and save when inserting the tool. When inserting the tool (for mostly the long-term samples), the tool must be made of materials the body can survive and accept.

Titanium, a good option sometimes, could be filled with problems if the doctor has to remove the bolt. Titanium metals have a bad result or effect in that they are not partially clear (like lightly frosted glass) to X-rays or MRI scans. So, when the doctors install it, it can hide body structure-related changes. Titanium is not an iron, so the magnet field used in MRI machines will not put into action a force on them. At the vertebra, the whole screw device, made up of smaller parts and rods, is titanium. The titanium bar, the simplest part, has a bend fit machine in one side. The doctor attaches the bar to two bolts, then attaches locking to both sides. The bend fittings let the doctor twist the bar, which can be a curve shape, to match the body structure of the spine.

5.6 Materials and methods

Doctors inserted two bolts carrying bar clamps into the handle of each backbone. Two bars were fixed in the handle, so the bar is parallel on both sides of the vertebra. The doctor then set the bar in its place. After several months, the actual bone and joints grew and expanded together, making them one part. The capability to bend and the reduction of pain relies on the ability of the patient. These steps only enclose movement between several of backbones, so they do not freeze the spine. The bar and bolts, after gripping the backbones in a fixed position of the spine, were made ready for use. Most surgeons prefer Titanium material because of its biocompatibility.

There are many types of spine surgery instrumentation: posterior lumbar cage, pedicle screws, and interior body as well cage. Pedicle screws are useful for catching the vertebral segment and stopping the motion of the segments.

Anterior cages are tools that, through an anterior method, are made to be constructed within the disk room. Allograft bone, titanium, or carbon may be made from them. Posterior cages are often planned to be constructed inside the disk gap but are modified to be mounted on the rear side. They can be built or made from the same materials as an anterior cage. Pedicle screws, which are usually used for spinal fusion surgery, give a means of catching the segments. These screws do not always make fixation for the spinal segments but act as fixed anchors that may be merged with the rod as demonstrated in Fig. 5.6. Two or three consecutive segments supply the screws, and a short rod is then used to merge the screws together. This arrangement prevents the segments from moving at the point of merging.

The bone is also growing; for this stability, the screws and rods are no longer required and can be kindly removed with a back operation. Most doctors, however, do not recommend this removal unless the pedicle screws cause the patient to feel pain. The use of pedicle screws has enhanced spinal fusion in the posterior lateral fusion with operation success rates of 60%–90%. Many surgeons understand that pedicle screws facilitate the recovery of patient health because they provide the spine and the segments with fast stability.

Interbody cages for spine fusion surgery: The cages are permeable for bone graft to grow from the segments of the body through the cage and into the other vertebral structure. The cages provide perfect fixation for the spine, so most patients do not need additional instruments or postsurgical posterior braces for support. Most of the cages are provided in the front side of the spine. The cages can be placed through a small slit or with a scope that allows the surgery to be done through several 1 in. slit. Actually, the most popular approach for placing titanium cages is toward a small incision as the endoscopic approach is hard and does not give perfect display.

Fig. 5.6 Pedicle screws [9].

In the past, spine fusions were all done with the bone of the patient. The amount of bone needed to be cropped from body bones in general, especially from iliac crest, was strongly reduced because only the inner spongy bone is needed for the surgery. A structural cortical of the spine fusion that describe the outer layer of the bone, was not recommended, and the outer part of the pelvis was not broken. Nowadays, there are a lot of bone graft replacements that may even cancel the necessity for bone graft crops. After an initial swell in publicity, it was shown that the passed cages did not correct the spine enough in many cases. At L4–L5 and up, specifically in patients with a wide disk gap, cages unfortunately do not serve as perfect as fixation. They are not acceptable in multilevel form in spine fusions with no supplemental fixation. There are various great characteristics for X-rays as well; most of the cages now are carbon fiber, which allows for bone graft recovery but does not serve as good fixation. In general, pedicle screw supplementation is necessary for most surgeries that need fixation for the spine; it is high quality, and most operations succeed because of them. Another design that is now available on the market is a metallic cage that can promptly hold and more tightly merge the anatomy the disk gap, considered to be part of a more recent design. That is because it does not have a cylindrical shape; it could be used in disk gaps when they are fixed tall. The cylinder structure needs to be quite big in a tall type of disk gap for the powerful subchondral bone. There is not yet enough previous experience with this kind of implant to recognize if it will work as a standalone device or not, but if physicians and researchers reach the point that they can use it alone, it will be very effective and helpful for the patient; and it will save recovery time (Fig. 5.7).

Cages risks and potential complications: Placing a cage could cause possible risks and issues. Breakage or scratching of any of the cages would be highly unlikely as they are stiffer than the vertebral segments. Taking them out is a possibility, especially for the titanium cylindrical cages instrument; the risk is low because of its biocompatibility.

Bone stimulators: Bone stimulators, which produce small electrical current into the bones especially the vertebral segments, can be used to serve and stimulate bone growth and enhance the spinal fusion. Bone stimulators can be implanted in the body or on the outside of the spine structure. Internal bone stimulators must have a battery that is provided under the derma, including wires that exist on top of the muscle placed in the posterolateral channel.

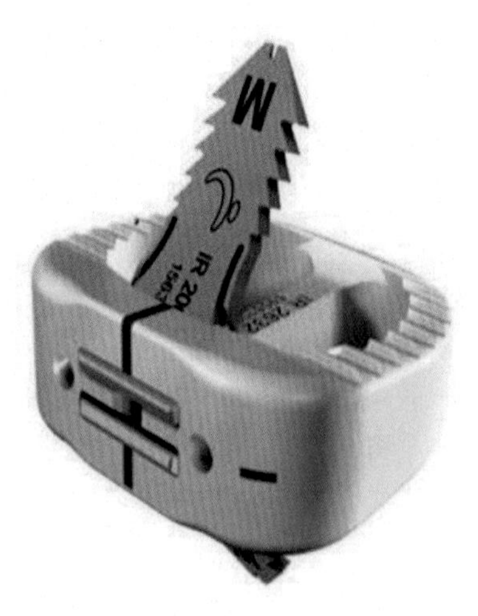

Fig. 5.7 Metal cage.

Bone graft for spinal fusion: Here, there is a problem in the disk between the vertebrae, and this depends on the situation; the doctors may remove the entire disk or part of it, and this process is called spinal fusion. To do this, the doctors insert a solid bridge, which is usually formed from titanium or titanium alloys, between the vertebrae to prevent the movement caused by them. At the design phase of the bridge, we considered it a good environment for the bone and the sizes to be suitable for the human body. It is impossible for the bone graft to form at the same time of the surgery, as we made the graft with a high characteristic and great environment, so with a period of time, the body will grow a new bone around the bridge. After a period of time, bone will be formed into one long bone or one vertebra; the instrumentation used by doctors, like screws and rods for long bone formation, are given and provide the stability for the section. Then, after some months, the bone will grow and heal as shown in Fig. 5.8. It shows the removable disk between L4 and L5, the lower vertebra and the place of the graft that the doctor will insert it.

Other ways to use a human bone instead of a bridge include using the patient's own bone, cadaver bone, or bone morphogenetic protein (BMP). As it is has been mentioned, there are many types of bones that can be used in this situation, but there are several considerations or factors to consider, including the number of levels of the spine involved, location of the fusion, patient risk and patient's personal situations, and the surgeon's performance.

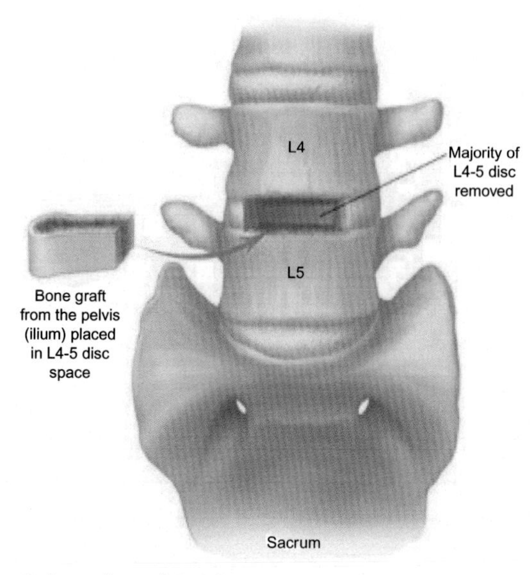

Fig. 5.8 Bone graft from the pelvis [9].

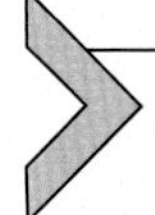

5.7 Discussion and results

5.7.1 Findings and discussion

Spinal stenosis is a rare complication, and in the field of spinal surgery, many decompressive procedures for this condition have become more common. Bone spurs on the vertebral bodies, posterior protrusion of the intervertebral disks, ligamentum flavum hypertrophy, and osteophyte formation are pathologic shifts. The standard approach for degenerative lumbar spinal stenosis is currently complete laminectomy with a success rate reportedly as high as 85%–90% [10, 11]. The clinical efficacy of a laminectomy, however, decreases over time [12]. Numerous experiments have shown that, unless fusion is carried out, complete laminectomy causes segmental instability. In 67 clinical cases, Postacchini et al. [11] compared multiple laminotomy and complete laminectomy cases. Compared to those who underwent complete laminectomy, postoperative vertebral instability was rare in patients undergoing multiple laminotomy. As the rate of decompressive surgery rises for this condition, incorporating full canal and foraminal decompression with limited resection of vital bony structures and supporting ligaments would be the optimal treatment (Table 5.1).

Table 5.1 Comparative result analysis.

	Intact (mm)		Bilateral laminotomy (mm)		Laminectomy (mm)	
Specimens	Flexion	Extension	Flexion	Extension	Flexion	Extension
Specimen 1	−1.118	0.944	−1.135	1.276	−2.015	1.187
Specimen 2	−1.116	0.992	−1.012	0.951	−1.433	0.860
Specimen 3	−1.123	1.033	−1.279	1.040	−1.338	0.938
Specimen 4	−0.892	0.710	−0.998	0.835	−1.402	0.944
Specimen 5	−1.004	0.829	−1.165	0.911	−1.410	1.155
Specimen 6	−0.0832	0.955	−1.196	1.286	−1.675	1.023
Specimen 7	−1.238	0.873	−0.856	1.115	−1.864	0.968
Specimen 8	−0.984	0.892	−1.243	0.969	−1.473	0.815
Ave.	−1.038	0.904	−1.111	1.048	−1.576	0.986
S.D.	0.126	0.095	0.134	0.155	0.232	0.122

L3–L4 segment intervertebral distortion on extension and flexion motions (8400 N mm): Intervertebral distortion did not show significant differences between three different decompressive procedures under extension motion ($P > .05$). However, intervertebral displacement was statistically higher in the laminectomy community under the flexion motion than in the intact or bilateral laminotomy groups ($P < .05$), as shown in Fig. 5.9.

Most decompression surgical methods include excision of the complexes of the interspinous or supraspinous ligament, altering an already pathologic biomechanical setting. Goel et al. [13, 14] found that, under normal conditions, when subjected to an external flexion moment around an anatomical segment, the supraspinous ligament was given the greatest force. Kanayama et al. [15] observed similar results and proposed that the paraspinal musculature provides compensatory flexibility in regions lacking this ligamentous support. Resection of sections or all spinal procedures, interspinal ligaments, and supraspinal ligaments, as well as iatrogenic paraspinal musculature injury, increases the volume of dead space excessively [16]. Using bilateral laminotomy instead of laminectomy, dead space and its consequent hazards are greatly reduced. The standard posterior median furrow, which would be lost using other more intrusive methods, are preserved by protection of the spinous processes and inter/supraspinous ligamentous complex.

In this analysis, as applied loading increases, the applied moment increases. The maximum moment was set at 8400 N mm, which, when the moment reached the preset value of 8400 N mm, stopped the flexion or extension motion. The facet joints locked and prevented posterior vertebral displacement under extension motion, which caused the moment to

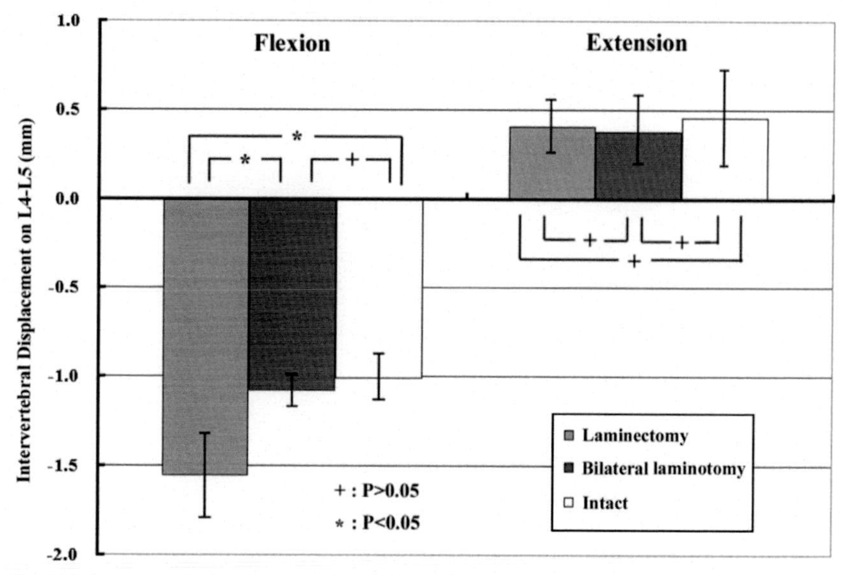

Fig. 5.9 Intervertebral displacement on L4–L5 (mm).

rapidly increase to the end point of 8400 N mm. In all three classes, this phenomenon was found. However, intervertebral displacement of the decompressive segment (L4–L5) was statistically greater in the laminectomy community under flexion motion than in the bilateral laminotomy or intact groups. This potentially causes the greatest stress on the decompressive segment and, despite protection of the facet joint, contributes to the degenerative acceleration of the decompressive segment following laminectomy. The present findings indicate that after a decompressive procedure, deconstruction of the anchor point for the supraspinous ligament may cause severe segmental instability. The findings also indicate that in lumbar spine decompressive surgery, bilateral laminotomy is generally successful.

5.8 Software

In the Software section, we used programs to simulate the spinal cord and apply some loads to check the patient's situation before the surgery; this step is very helpful to decrease the side effects that may occur after surgery.

Firstly, we used COMSOL Multiphysics 5.3a, which is a program for analyzing information by simulating them. COMSOL provides a good test and analysis for mechanics, which can be useful in geometric applications. Fig. 5.10 shows the user interface of COMSOL.

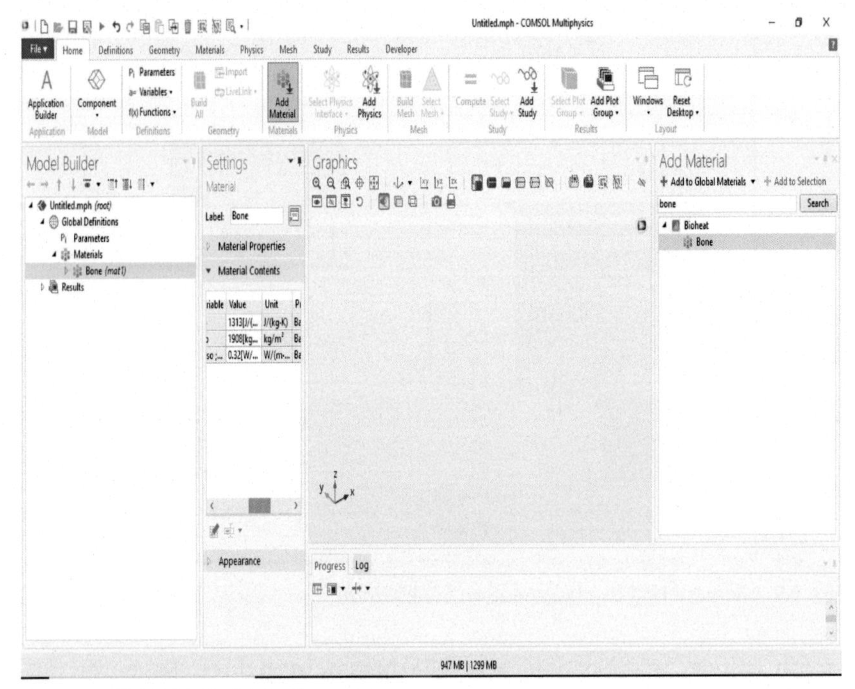

Fig. 5.10 COMSOL Multiphysics program.

As shown in Fig. 5.10, this software is useful for applying various loads that exist to examine the spine in different situations. Fig. 5.11 shows the vertebra we designed in specific dimensions. We were able to complete the spine for the lower back section, which will be examined after applying the spacers to it, to show the results and the amount of displacement caused by these kinds of loads.

These results could help a surgeon to decide whether to authorize the surgery or not, corresponding to these data.

5.9 Hardware

In the Hardware section, we used a spinal cord model to explain the whole procedure. This model contains 5–6 vertebrae, and the main group is the lumbar spine group. Ligaments and nerves are also shown in our model (Fig. 5.12).

The price of the model is $38 (USD) and is available via the Amazon website.

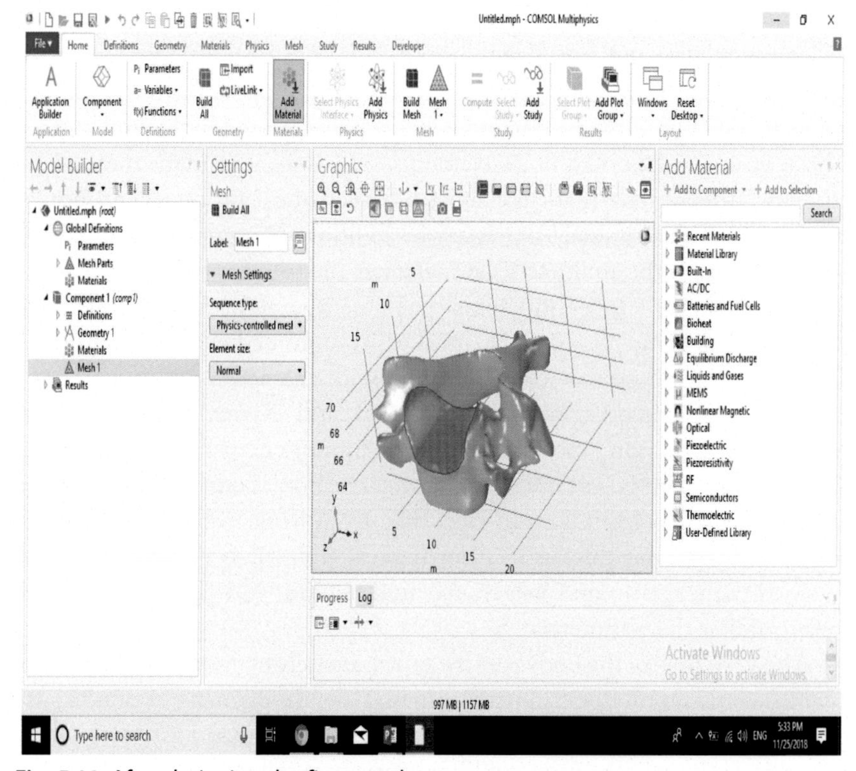

Fig. 5.11 After designing the first vertebra.

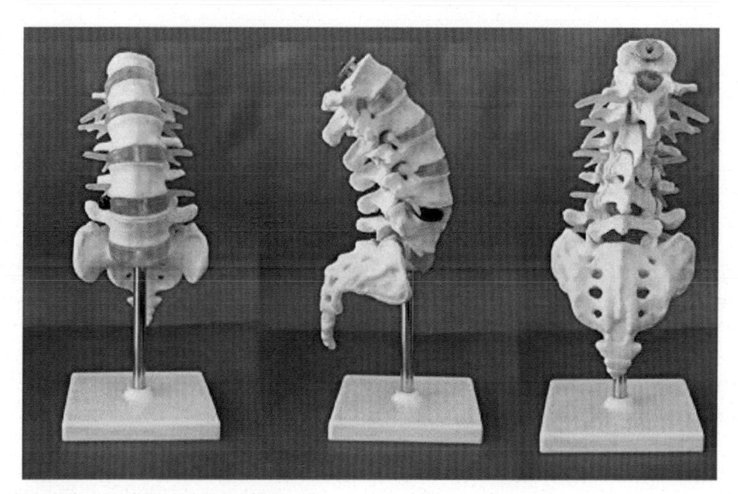

Fig. 5.12 Spinal cord models [17].

We also utilized the brace in the Hardware section, which gives mechanical support after the laminectomy or laminotomy. As mentioned in the Materials section, titanium gives the mechanical support for the spine in a spinal fusion, but it is very expensive, hence it is replaced with stainless steel. Numerical simulation of finite elements (FE) is an important method for studying phenomena that, like most biomechanical processes, cannot be processed by experimental strategies. In addition, computational modeling approaches have the ability to scale back rates and prevent wasted time in the event of the latest successful spinal pain management. We designed our model with specific pure mathematical expressions deduced from surgeons' data. We tend to use materials supported with characteristics of vertebra.

Firstly, utilizing lateral imaging (MRI and X-ray pictures) of the lumbosacral junction, the body trunk measurement is determined using the delineate of body part spinal curvature. The bony body of every vertebra is large and wider from facet to facet than from front to back, as shown in Fig. 5.13, and a bit thicker in front than in back. It is planar, or slightly recessed higher, than and below, recessed behind, and deeply constricted before and at the perimeters.

The vertebrae of the body part are approximately cylindrical with a lateral (width) diameter of 40–50 mm and a mesial (depth) diameter of 30–35 mm. The vertebrae of the body part are thicker anteriorly than posteriorly, contributing to the spine's anterior gibbose curvature called the spinal curvature of the body part. The heights of the dorsal body are therefore 25–27 mm,

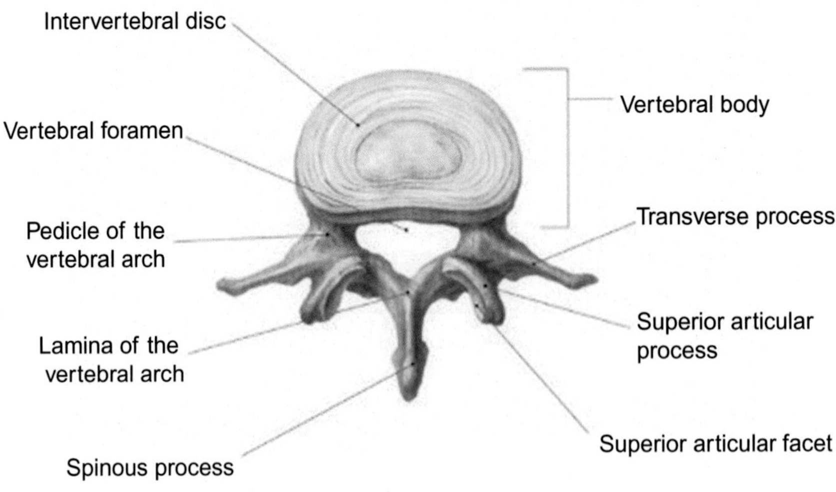

Fig. 5.13 Body part vertebrae [17].

while the heights of the ventral body are 26–29 mm. The boney body is composed of an associated outer layer of high-strength animal tissue bone that is internally reinforced by the cellular bone as a network of trabeculae, known as vertical and horizontal slender bone, struts. As shown in Fig. 5.14, the upper and lower surface of the bony body is covered by the bony endplates of skinny plant tissue bone perforated by several small holes that allow the passage of bone metabolites to the central regions of the avascular disks.

A trial of stout bone pillars, known as the pedicles, supporting the posterior components may be at the higher end of the posterior surface of the bone body. A skew bone plate, known as the plate, goes to the sheet from every pedicle wherever they combine. A channel encloses the pedicles and laminae along the arch with the posterior surface of the bone body: the bone gap. The pointing method forms the peak-like junction of the two laminae within the sheet, while an extended planar bar of bone, known as the process, begins at the lateral junction of each plate and pedicle. At the base of each operation, two bone extensions rise from the lamina: the superior upwards, the inferior articular method downwards.

Following the anatomical dimensions of the associated vertebrae, the bone disks distinguish the neighboring vertebrae, approximately cylindrical with a lateral diameter (width) of 40–45 mm and a mesial diameter (depth) of 35–40 mm. Within the body part area, the quantitative relationship between

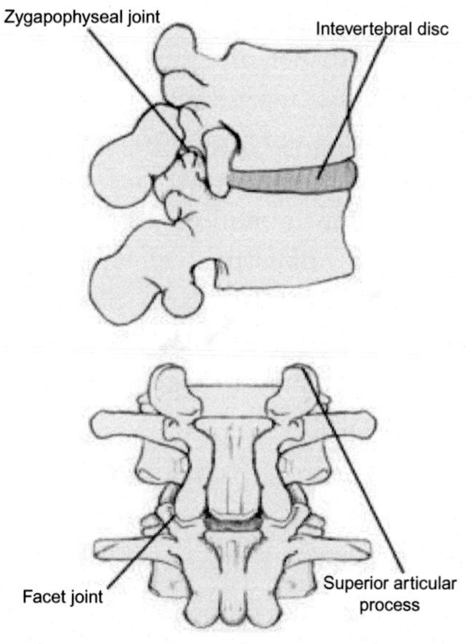

Fig. 5.14 Vertebra in different angles [18].

the height of the disk and the height of the bone body is 1:3, and the height of the body part is 10 mm. Once the disks are below the compressive load, water is drained out of them during everyday operation so that they lose their height. Water falls through the disks once at bedtime, so that the height of them is set once more.

Intervertebral disk tissue is anisotropic in its composition. The disk consists of three components: the jellylike core; the pulposus nucleus, enclosed by the annulus fibrosus' concentrically arranged fibrous layers, or lamellae; and the superior and inferior rubber endplates. Seven ligament types are distinguished in the body portion of the spine; five of them link the many elements of the posterior vertebrae, and two of them connect the bone bodies themselves.

The lower and higher ends of the inner surfaces of the adjacent laminae are bound by the ligament flavum (LF), the foremost elastic ligament of the spine, closing the distance between the successive laminae. Intertransverse ligaments (ITL) bind thin sheets of scleroprotein fibers to cross-sectional processes. Interspinous ligaments (ISL) bind the opposite edges of scleroprotein fibers to acanthoid processes, while supraspinous ligaments (SSL) connect connective tissue fibers to the peaks of adjacent acanthoid processes. The capsular ligaments (CL) link the circumferences of the articular side joints of the connection, being perpendicular to the joints' surface. The anterior longitudinal ligament (ALL) protects the bone body and disk anterior surfaces, which are firmly attached to the bone and delicate to the disks. Consequently, the bone is broader, and the disks are smaller. The posterior longitudinal ligaments (PLL) comprise the posterior parts of the disks and bone bodies, which are firmly connected to the disks and delicate to the bone. Consequently, the disks are broader, and the bone is shorter.

And at microscopic strains, nonlinear elastic materials have nonlinear stress-strain relationships as opposing hyperplastic materials wherever stress-strain curves at moderate to giant strains become considerably nonlinear. For modeling metals, various ductile materials, and nonlinear models of soils (such as the Duncan-Chang model), Ramberg-Osgood materials are required for this group.

5.10 Conclusion

Our study consists of solving the surgical mistakes that surgeons make before they occur and possibly cause harm to patients. In this study, we were able to solve many other issues in different parts of the body and simulate them using the COMSOL Multiphysics program.

FE Mechanical simulation is a good software for studying an issue that cannot be executed successfully in the laboratory. Also, mechanical simulation methods have the ability to scale back prices and to reduce execution time throughout the event of recent spinal examinations.

References

[1] K. Young-Hing, W.H. Kirkaldy-Willis, The photophysiology of degenerative disease of the lumbar spine, Orthop. Clin. North. Am. 14 (1983) 491–504.

[2] K. Sasaki, Magnetic resonance imaging findings of the lumbar root pathway in patients over 50 years old, Eur. Spine J. J4 (1995) 71–76.

[3] S. Etebar, D.W. Cahill, Risk factors for adjacent segment failure following lumbar fixation with rigid instrumentation for degenerative instability, J. Neurosurg. 90 (1999) 163–169.

[4] O.L. Lai, L.H. Chen, C.C. Niu, T.S. Fu, W.J. Chen, Relation between laminectomy and development of adjacent segment instability after lumbar fusion with pedicle fixation, Spine (Phila Pa 1976) 29 (2004) 2527–2532.

[5] M. Lee, R.J. Bransford, C. Bellabarba, J.R. Chapman, A.M. Cohen, et al., The effect of bilateral laminotomy versus laminectomy on the motion and stiffness of the human lumbar spine, a biomechanical comparison, Spine (Phila Pa 1976) 35 (2010) 1789–1793.

[6] https://www.comsol.com/. (Retrieved 8 January 2021).

[7] (a) J. Englund, Lumbar spinal stenosis, Curr. Sports Med. Rep. 6 (2007) 50–55. (b) E. Arbit, S. Pannullo, Lumbar stenosis: a clinical review, Clin. Orthop. Relat. Res. (2001) 137–143.

[8] S. Hall, J.D. Bartleson, B.M. Onofrio, Lumbar spinal stenosis: clinical features, diagnostic procedures, and results of surgical treatment in 68 patients, Ann. Intern. Med. 103 (1985) 271–275.

[9] https://www.spine-health.com/treatment/spinal-fusion/pedicle-screws-spine-fusion. (Retrieved 5 January 2021).

[10] M.W. Fox, B.M. Onofrio, A.D. Hanssen, Clinical outcomes and radiological instability following decompressive lumbar laminectomy for degenerative spinal stenosis: a comparison of patients undergoing concomitant arthrodesis versus decompression alone, J. Neurosurg. 85 (5) (1996) 793–802.

[11] F. Postacchini, G. Cinotti, D. Perugia, S. Gumina, The surgical treatment of central lumber stenosis. Multiple laminotomy compared with total laminectomy, J. Bone Joint Surg. (Br.) 75 (3) (1993) 386–392.

[12] F.B. Ensink, P.M. Saur, K. Frese, D. Seeger, J. Hildebrandt, Lumbar range of motion: influence of time of day and individual factors on measurements, Spine 21 (11) (1996) 1339–1343, https://doi.org/10.1097/00007632-199606010-00012.

[13] V.K. Goel, S.J. Fromknecht, K. Nishiyama, J. Weinstein, Y.K. Liu, The role of the lumbar spinal elements in flexion, Spine 10 (6) (1985) 516–523, https://doi.org/10.1097/00007632-198507000-00005.

[14] V.K. Goel, S. Goyal, C. Clark, K. Nishiyama, T. Nye, Kinematics of the whole lumbar spine: effect of discectomy, Spine 10 (6) (1985) 543–554, https://doi.org/10.1097/00007632-198507000-00008.

[15] M. Kanayama, K. Abumi, K. Kaneda, S. Tadano, T. Ukai, Phase lag of the intersegmental motion in flexion-extension of the lumbar and lumbosacral spine: an

in vivo study, Spine 21 (12) (1996) 1416–1422, https://doi.org/10.1097/00007632-199606150-00004.
[16] T.G. Mayer, H. Vanharanta, R.J. Gatchel, V. Mooney, D. Barnes, L. Judge, S. Smith, A. Terry, Comparison of CT scan muscle measurements and isokinetic trunk strength in postoperative patients, Spine 14 (1) (1989) 33–36, https://doi.org/10.1097/00007632-198901000-00006.
[17] https://www.encoris.com/spine-model/. (Retrieved 10 January 2021).
[18] https://emedicine.medscape.com/article/1899031-overview. (Retrieved 10 January 2021).

CHAPTER SIX

Construction of an ultrasonic sight device for visually impaired people

Dilber Uzun Ozsahin[a,b,c], John Bush Idoko[d], Abdullah A. Usman[b], Rehemah Namatovu[b], Kamil A. Ibrahim[d], and Ilker Ozsahin[a,b,e]
[a]DESAM Research Institute, Near East University, Nicosia, Turkish Republic of Northern Cyprus, Turkey
[b]Department of Biomedical Engineering, Near East University, Nicosia, Turkish Republic of Northern Cyprus, Turkey
[c]Medical Diagnostic Imaging Department, College of Health Sciences, University of Sharjah, Sharjah, United Arab Emirates
[d]Applied Artificial Intelligence Research Centre, Department of Computer Engineering, Near East University, Nicosia, Turkish Republic of Northern Cyprus, Turkey
[e]Brain Health Imaging Institute, Department of Radiology, Weill Cornell Medicine, New York, NY, United States

6.1 Introduction

Blindness is a condition of missing the visual discernment due to physiological or neurological variables in part or in full. The main concept here is to provide electronic assistance as direction in overcoming the visualization control problem by proposing a basic, effective, configurable electronic direction framework for visually impaired individuals [1]. The human sight system is among the complex systems in the central nervous system. The visual framework incorporates the eyes, the interfacing pathways through the visual cortex, and other parts of the brain that receive the reflected light from the surrounding objects and form the image [2]. The presence of any kind of defect in the visual system can cause visual impairment, and in serious cases, blindness can occur with total damage to the visual system. It is evaluated that over 285 million individuals are outwardly impeded around the world with 39 million being dazzle and the rest having obscured dreams. It is additionally assessed that an expansive extent of the daze individuals (approximately 90%) live in developing nations, and most of them cannot travel without any kind of assistance.

Most of the visual deficiency is basically due to birth shortcomings and cannot be redressed at the displayed time. Dazzle individuals endure from numerous drawbacks. Since they cannot see, they neither study nor draw

Modern Practical Healthcare Issues in Biomedical Instrumentation
https://doi.org/10.1016/B978-0-323-85413-9.00010-4

like the typical individuals, and they cannot travel or explore effortlessly without the assistance of a few kinds of external tools, such as a guiding pet, cane, and so on. Ultrasonic amplifying is the procedure utilized by bats and various other living things for navigational purposes. In order to mimic how it works, scientists effectively duplicate a few parts of it and propagate it to an awesome degree. The creation and application of this system has an old history since the postwar period. Although this system is applied to replace a guide dog or cane, more recent efforts have been dedicated to the devices and the system designed to support the movement. The ultrasonic haptic vision framework enables a person to investigate passages and around huge objects without vision by utilizing an ultrasonic rangefinder [3]. The thought behind this amplifies to create a sixth substantial system that works with the body in a common and client-focused plan and engages the client to see without vision. The proposed device is specifically designed for visually impaired people. Here, the main idea is to produce a reliable and fully functioning device that will aid in the movement of blind people at a low cost and improve their quality of life. The system uses Arduino, and with the help of a mini-vibrating disk and a speaker, the device vibrates when there is an object in front of the person, thereby issuing an audible alert.

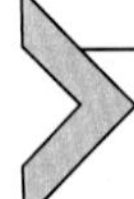

6.2 Blindness

6.2.1 Types of blindness and causes

Glaucoma: This occurs when eye fluid is not drained properly, causing pressure built in the optic nerve. It can cause blindness if not treated at the early stage.

Diabetic retinopathy: This condition is caused by high blood sugar and high blood pressure, which in turn causes damage of blood vessels in the optic nerve, leading to leakage of fluid. It has three stages and affects the central part of the retina.

Cataracts: These are caused by build-up of protein that cloud the eye lens. These protein patches continue to develop either from the front side or the back of the lens. They usually appear due to old age. Statistically, about 12 million people are blind due to cataracts.

Optic neuritis: Optic neuritis is a result of the aggravation of the optic nerve. It can result in sight misfortune, which may affect one or both of the eyes. It may not continuously require treatment and can regain sight without further treatment. Solutions such as corticosteroids, can offer assistance to speed up recuperation.

Trachoma: This is caused by chlamydia trachomatis and spreads through contact with flies or an infected person. First, the eye is inflamed, causing scar tissue build-up leading to the eyelid and making it tight, pulling eyelashes inwards; this, in turn, leads to cornea inflammation.

Refractive error: It is caused by irregularities in eye shape, which include myopia (short-sightedness), hyperopia (long-sightedness), and astigmatism (curved cornea). This causes image focus to be difficult, leading to blurred vision.

River blindness: This is caused by parasitic flies living around water bodies. When nibbled by the flies, the body is attacked by worm hatchlings that become worms that can live for a long time. Female worms produce microfilariae. This spreads all over the body and can be transmitted to others. When the microfilaria dies, they can cause a response that leads to a gigantic disturbance, aggravation, and tingling. If the hatchlings travel to the eyes, they can cause irreversible damage.

Pink eye: This is also known as conjunctivitis. It is an inflammation of the conjunctiva; that is the straightforward layer that shields the white part of the eye and the internal surface of the eyelids. It makes the eye assume pink or red color. Pink eye is commonly caused by viruses, microbes, or aggravation (Table 6.1).

6.3 Blindness preventable

Blindness can be avoided and prevented through access to proper medical care and education. Almost all trauma that leads to blindness can be stopped via eye protection. Lack of proper nutrition that leads to blindness can be avoided with the appropriate diet. Sometimes the causes of visual impairment via glaucoma can be avoided through early detection and proper treatments. Blindness and visual impairment lead by infectious pathogens have been decreasing via international public health measures. Blindness from diabetic retinopathy is avoidable via steady control of blood sugar level, exercise, prevention of smoking and obesity, and a nutritious diet. Lately, there has been a rise in the population of visually impaired people from situations that are a result of old age. As several people in the world experience longevity, it will lead to more blindness from infections such as macular degeneration. Regular doctor checkups with eye examinations can uncover a preventable blinding disease before any visual loss.

Patients that have incurable blindness require tools and assistance to reorganize their environment and the way they handle their everyday tasks.

Table 6.1 Number of visual impaired people [4].

WHO region	Total population (millions)	Blindness No. in millions (percentage)	Low vision No. in millions (percentage)	Visual impairment No. in millions (percentage)
Africa	804.9 (11.9)	5.888 (15)	20.407 (8.3)	26.295 (9.2)
America	915 (13.6)	3.211 (8)	23.401 (9.5)	26.612 (9.3)
EMR (Eastern Mediterranean Region)	580.2 (8.6)	4.918 (12.5)	18.581 (7.6)	23.499 (8.2)
Europe	889.2 (13.2)	2.713.(7)	25.502 (10.4)	28.215 (9.9)
SEAR (South-East Asia Region) (India excluded)	579.1 (8.6)	3.974 (10.1)	23.938 (9.7)	27.913 (9.8)
WPR (Western Pacific Region) (China excluded)	442.3 (6.6)	2.338 (6)	12.386 (5)	14.724 (5.2)
India	1181.4 (17.5)	8.075 (20.5)	54.544 (22.2)	62.619 (21.9)
China	1344.9 (20)	8.248 (20.9)	67.264 (27.3)	75.512 (26.5)
World	**6737.5 (100)**	**39.365 (100)**	**246.024 (100)**	**285.389 (100)**

Some organizations offer support and helpful resources to people suffering blindness. Such support includes visual aid and text reading software with other simple and complex techs to aid people with highly compromised visibility to function more effectively. In the United States and other developed countries, financial aid via various agencies pays for the support and training necessary to enable the ability of visually impaired people to function effectively with and without assistance.

6.4 Other devices that aid blind people

A wearable walking guide system in South Korea is currently under creation. It will be produced to notice objects around the person using ultrasound sensors and to alert the data-using bone conduction vibrator with a beep sound and TTS voice massage [5]. A pool digital assistant (PDA) is utilized as a fundamental controller and gives the caution alert. It sends data sound to the daze, and voice acknowledgment strategy is utilized for the client comfort. AI glasses could be a gadget that mixes computational geometry, artificial insights, and ultrasound methods to form valuable assistance for the outwardly impeded. The model links glasses with stereo sound sensors and GPS innovation connected to a device. This gives spoken headings and recognizes groups of money, perused signs, and colors. It moreover applies machine learning to recognize distinctive places and obstacles. Since it utilizes ultrasound technology, it can distinguish translucent impediments, like glass walls.

6.5 Working principle

This device works using ultrasonic sensor and Arduino as the main components. The ultrasonic sensor is connected to the breadboard. In this study, an HC SR04 sensor was utilized. It has four pins: Ground, VCC, Trig, and Echo. These four pins are connected to the breadboard and are internally connected to the Arduino using jump wires. One end of the wires is connected to the breadboard placed next to the sensor's pin with the other end connected to the Arduino and the buzzer connected to the breadboard. The buzzer has two legs; the little leg (negative) and the long leg (positive). These legs are connected separately to the Arduino using jump wires. This ensures that the information from the Arduino is passed along to the buzzer. A terminal of the speaker is connected to the breadboard while the other terminal is connected to the Arduino with the help of jump wires. The Arduino must be programmed to perform the necessary tasks of the device. This means that the Arduino needs a code that will be uploaded using the Arduino app from the computer [5]. The Arduino also needs power supply using a battery or USB cable. Finally, the device is packaged so that it can be comfortably held by the user. The workflow and full circuit connection within the device are depicted in Figs. 6.1 and 6.2 respectively.

6.6 Results and discussion

All the components of the gadget cost less than $35 (USD) and are easily accessible. The proposed device encompasses a wide run location of up to 450 cm and a location exactness of 100 cm for moving objects. The Ultrasonic Sight device concept is exceptionally basic and cost effective. The transmitter sends the sonar signal to the impediments and subsequently sends them back; afterwards, the reflected flag is delivered to the recipient.

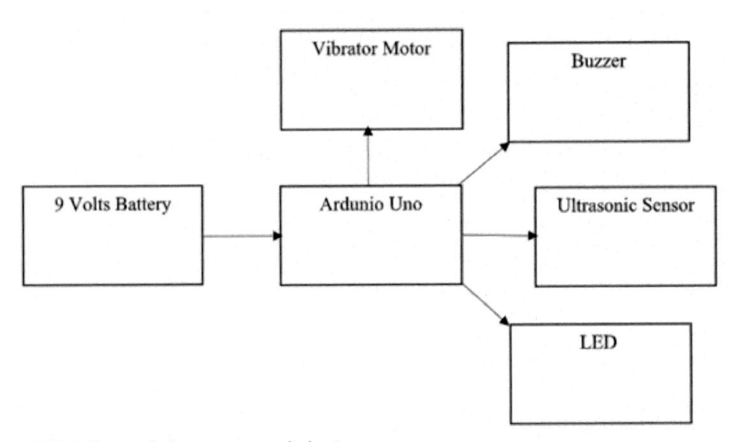

Fig. 6.1 Workflow of the proposed device.

Fig. 6.2 Full circuit connection of the device.

The microcontroller of the framework, the Arduino Uno, assesses the sound to recognize the nearness of any impediment [6]. The device is secured because it does not have any contact with the objects when deciding the nearness of the impediment. This procedure demonstrates that sonar innovation is exceptionally appropriate for utilization in assistive devices for impediment evasion systems.

The proposed device was tested on a series of objects (input), and the recognition rate with the alerting technology was as accurate as expected. The gadget range and exactness enable an individual with no vision to identify objects with the interval of $0.5\,m^2$. To exclude head-level impediments, the client must clear the head region along the target location. Amid the tests, it was moreover found that it is conceivable to track a moving impediment in the range of 100 cm. The preparatory examination of the gadget was determined to be a victory based on the criticism received from the volunteers. All the invited volunteers amid the tests realized the adequacy of the haptic feedback and the instinctive learning and adjustment forms the ease of use of the gadget. The device produces a quick reaction and enables an individual to be aware of moving objects as well as inactive impediments.

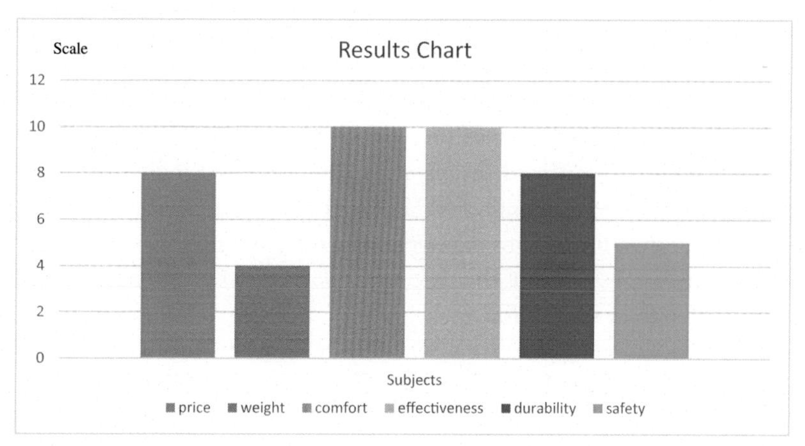

Fig. 6.3 Results chart.

Fig. 6.3 shows the assessment from the volunteers, and it shows how concerned the volunteers were about the comfort of the user and the effectiveness of the assistive device. Other features, such as durability and price, had the same number of votes from the volunteers. The weight and safety were voted 4 and 5 respectively out of the scale of 12. The rates of volunteers who were sure to distinguish impediments underneath and over the abdomen were the same. The number of participants who were certain to identify a moving impediment utilizing the gadget was higher than the rate for those using the gadget in identifying and locating stairways. In addition, a large part of the participants was sure to utilize the gadget since it is able to distinguish distinctive sorts of deterrents at distinctive levels, impediments which are underneath abdomen level, over abdomen level, head-level, and stairways.

Some of the pros of the proposed device includes that it easily detects objects; works under dark environments; is low cost, resistant to dirt, dust, and high moisture environment; and is easy to use. However, it has some cons which include limited detection range, inability to be used under water, and difficulty detecting objects with soft fabrics as cover.

6.7 Conclusion

The objective of this study was to provide a navigation aid to visually impaired people and to enhance their movement independently and safely. This procedure is basically used in aircraft navigation, and it has helped to reduce errors caused by the accelerometer and double integration. Additionally, the use of it is incredibly important because, without it, drift errors caused by double integration and the accelerometer would be relatively higher in magnitude and would lower the effective range of the electronic travel aid. All the users that tested the proposed device rated it above 85%. The ratings were relative to detection and alert-beeping accuracy. Unfortunately, it has some limitations relative to the ultrasound reflection properties that made it impossible for the device to detect some obstacles at certain angles with soft or small surfaces.

This device does not cover a wide range in terms of distance, and the direction is also limited. Further research could target the incorporation of the GPS system. The inclusion of the GPS system would introduce a tracker that uses the satellite network to locate the device and would also generate real-time alerts when it goes off.

References

[1] J. Wilson, Sensor Technology Handbook, Elsevier, Oxford, UK, 2005.
[2] S. Chaurasia, K.V.N. Kavitha, An electronic walking stick for blinds, in: International Conference on Information Communication and Embedded Systems (ICICES2014), Chennai, 2014, 2014, pp. 1–5, https://doi.org/10.1109/ICICES.2014.7033988.
[3] M.H. Wahab, A.A. Talib, H.A. Kadir, A. Noraziah, R.M. Sidek, Smart cane: assistive cane for visually-impaired people, IJCSI Int. J. Comput. Sci. Issues 8 (4) (2011). No 2.
[4] Estimating the number of visual impaired people. http://birdsnestfoundation.org/perfect-food-for-perfect-vision/. (Retrieved 20 November 2020).
[5] A. Sharma, R. Patidar, S. Mandovara, I. Rathod, Blind audio guidance system, in: National Conference on Machine Intelligence Research and Advancement (NCMIRA, 12), India, 2012.
[6] T. Mohammad, Using ultrasonic and infrared sensors for distance measurement, World Acad. Sci. Eng. Technol. 27 (2009).

CHAPTER SEVEN

Design and implementation of a smart stick for visually impaired people

Ramiz Musallam Salama[a], John Bush Idoko[b], Kevin Meck[c], Sunsley Tanaka Halimani[c], and Dilber Uzun Ozsahin[c,d,e]

[a]Department of Computer Engineering, Near East University, Nicosia, Turkish Republic of Northern Cyprus, Turkey
[b]Applied Artificial Intelligence Research Centre, Department of Computer Engineering, Near East University, Nicosia, Turkish Republic of Northern Cyprus, Turkey
[c]Department of Biomedical Engineering, Near East University, Nicosia, Turkish Republic of Northern Cyprus, Turkey
[d]DESAM Research Institute, Near East University, Nicosia, Turkish Republic of Northern Cyprus, Turkey
[e]Medical Diagnostic Imaging Department, College of Health Sciences, University of Sharjah, Sharjah, United Arab Emirates

7.1 Introduction

Vision is the most important part of human physiology as 83% of information human beings get from the environment is via sight. Visually impaired people are individuals who are partially or completely blind. In 2011, the World Health Organization (WHO) conducted a study in which it found that approximately 285 million people are visually impaired. Of those 285 million people, about 10% of them are completely blind with the rest of the 90% of the visually impaired population having partial blindness [1]. These affected populations face the day-to-day challenges of navigating their way around places whenever they are outdoors, and this requires some form of assistance. This study proposes a new technique for designing a smart stick to help visually impaired people using sensors and warnings that will provide a safer and easier way of navigation. The walking cane and/or walking stick is the conventional and archaic means of a navigation aid for people who are visually impaired. Dogs may also be trained and used to assist visually impaired people with navigation, but these aids come with many disadvantages. The disadvantages of these conventional aids are the requirement of training and skill for use; limited range of motion, and risk of inessential or inaccurate information given to the user. A white

Modern Practical Healthcare Issues in Biomedical Instrumentation
https://doi.org/10.1016/B978-0-323-85413-9.00006-2

cane or walking stick does not give information about the obstacles above knee level, and those objects that are at a distance of 1 m or more. Furthermore, the white stick gives a warning about 1 m before the obstacle; for a normal walking speed of 1.2 m/s, the time to react is very short (only 1 s). The stick cannot detect certain obstacles like rears of trucks, low branches, etc. Even though the guide dogs were the initial companion of the blind, newer technologies have been introduced, and these include adjustable walking sticks, elbow canes, etc. The research is a modification of this cane with some electronic components and sensors designed to solve some of these problems that come with conventional navigation aids. The ultrasonic sensors, water sensor, and RF transmitter/receiver are used to sense and record information when an obstacle is detected. The ultrasonic sensor has the capacity to detect an obstacle within the distance range of between 2 and 450 cm. When the sensor is triggered by an obstacle within this range, the user will be alerted. A water sensor is used to detect if there is any water present in the path of the user. The advantage of using ultrasound by blind guidance systems is that it is least affected by environmental noises. The recent advancement in hardware and software in modern technology has made it easier to provide a much-needed intelligent navigation system to the visually impaired [2–5]. Recent studies have mainly focused on the designing of an Electronic Travel Aid (ETA) to aid in the successful and free navigation of the blind. High-end technological solutions have also been employed to help the blind with easier ways of navigation without much assistance. Some inventions, however, are limited because they require a separate power supply or navigator, which makes the whole unit bulkier for the patient to carry.

Another advantage of using ultrasonic technology is that the technology is relatively cheap to acquire and to use. Moreover, ultrasound emitters and detectors are portable components that can be carried around easily and integrated within the normal walking stick. The RF module also helps the visually impaired person to locate the stick wherever it is kept. To locate the stick, the user will have to press a button on a remote control, which will trigger the buzzer. The user will then follow this sound in order to locate the stick [6, 7].

Vision is the most important part of human physiology as 83% of information human being gets from the environment is via sight, and moving through an unknown environment without any visual can be very difficult. Dynamic obstacles produce a sound when in motion, and blind people develop an advanced sense of hearing and touch to act as additional sensors to detect these obstacles. To correct these shortcomings, the smart electronic aid (the proposed novel technology) is designed in such a way that it includes an ultrasonic sensor for obstacle detection, supported with heat and water detection. In this system, vibratory motors are used to detect the moving obstacles [6]. The intensity of vibration depends on the speed of the moving obstacles.

7.1.1 Objective

The research here aims to provide a better alternative to the conventional walking stick, which, with newer technology, makes it easier for the visually impaired to detect obstacles more efficiently without much assistance. In a world where privacy and independence are now a necessary requirement and right, this stick comes in as a partial solution to help the visually impaired to become more independent when moving outdoors without seeking help from others; in public spaces, sometimes people with these disabilities may feel the need to seek help from others. Unfortunately, in most countries, visually impaired people have failed to fully integrate and become part of the society. The intelligent walking stick, built with the highest degree of precision, assists the blind individuals to move from one location to another without any assistance from anybody.

7.2 Proposed work

The ultrasonic sensor functions as human eyes in the suggested scheme; it sends ultrasonic waves and the waves repulse back. The sensor detects the barrier and the distance between the blind individual and the barriers as a result of this procedure and sends data to the microcontroller. The sound-based ultrasonic sensors are responsible for this [6]. The sound waves are then transmitted from the sensors to the obstacle which, with a resolution of 0.3 cm, can feel the distance up to 12 ft. The sensors are put in five places: the left, right, center, center left, center right, to cover the highest possible directions with minimal sensor use, as demonstrated in Figs. 7.1 and 7.2.

7.2.1 Components

This scheme is used to identify the barrier (if any) by the ultrasonic sensors. The sensors are set to identify barriers within the range of variance chosen by the individual. Obstacles discovered in various directions are stated for easy

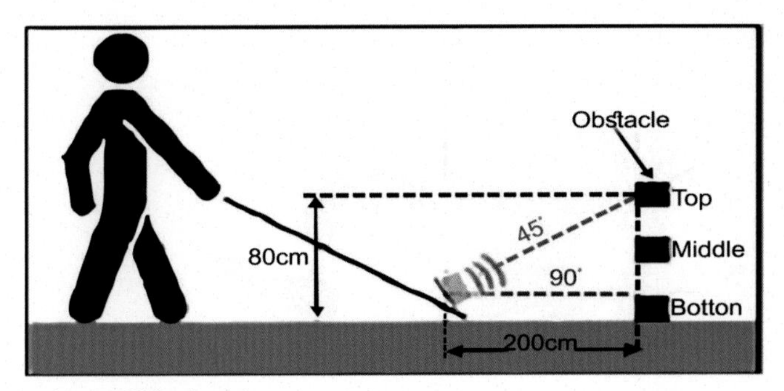

Fig. 7.1 Working of blind stick.

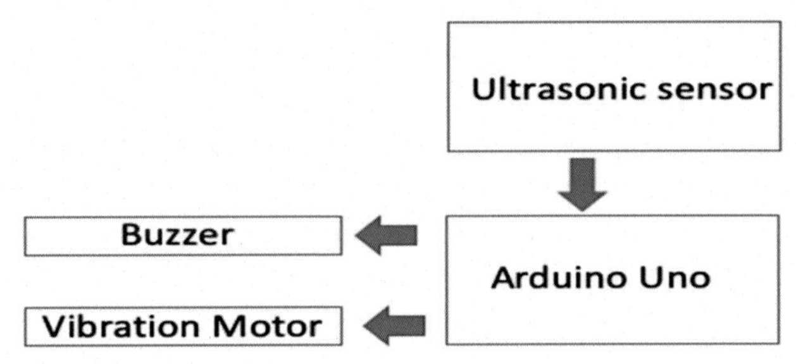

Fig. 7.2 Block diagram of the project.

identification with distinct beep patterns [8–11]. The ultrasonic sensors emit sound scopes that are inaudible to humans with ultrasonic spectrum frequencies (>20 kHz).

7.2.2 Working principle

The system's primary benefit is that it helps blind individuals with carefree navigation indoors and outdoors. The tools in the stick make it comfortable and simple to manage. The intelligent stick helps to detect barriers in front of the user at a distance. The system is appropriate for indoor as well as outdoor environments. Information about the obstacles is provided through the buzzer, eliminating the trouble of knowing the patterns of vibration used in previous schemes [12, 13]. The scheme is a cost effective, portable, and navigational aid for visually impaired individuals. The system's primary portion is the microcontroller that regulates other system parts.

7.2.3 Software requirements

The Arduino Uno is a simple microcontroller that has gained significant momentum in the technological world. The Arduino Uno is an open-source software, meaning the users can alter and use the software however they please. Arduino Integrated Development Environment (IDE) is an open-source software that is mainly used for writing and uploading programs and codes into Arduino-compatible modules. The architecture of a simple Arduino is depicted in Fig. 7.3.

The code and software used: Arduino consists of hardware circuit boards and programmable software. These circuit boards are able to read inputs and convert them into outputs using a set of instructions, or codes, predetermined by the user. Input devices can be a light, a sensor, a message, or a button that can be activated as an output. This requires a set of codes or programming language and Arduino IDE software. Arduino has been used in recent years for a number of scientific inventions like programming robotics, musical instruments, 3D printing, detectors for earthquakes, climate change, smoke, etc.

The coding language: The coding language for Arduino is a general purpose programming language called C++ with an augmentation of special functions and methods, and this will be further elaborated on in subsequent sections of the paper. When a "sketch" is created, which is the label given to the Arduino code files, processing and compilation into machine language is then executed. The Arduino IDE is the software program used for C++ code editing for Arduino. This is where the code is typed prior to uploading for the Arduino board to be programmed. Arduino code is also known as a sketch. The minimalist design of the IDE is very apparent. The menu bar has

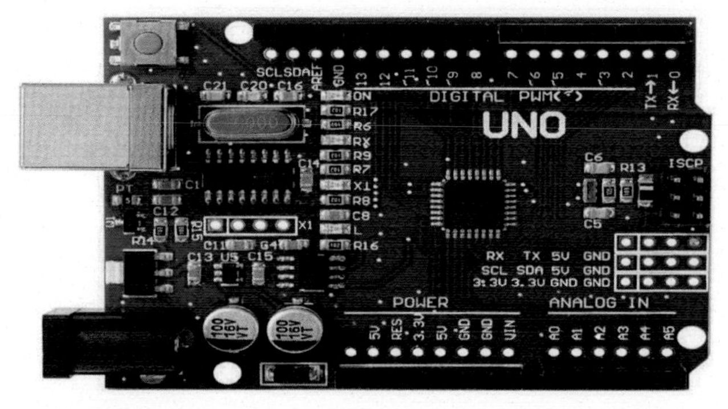

Fig. 7.3 Arduino Uno.

only 5 headings with buttons to verify and upload the sketches located in a series just below it. The purpose of the IDE is to translate and compile the sketches and put them into a code that is understandable to the Arduino. After compilation of the Arduino code or sketches, it is uploaded to the memory of the board. In the event that the Arduino code has some errors, an error message is flagged for proper debugging. The precise and strict syntax requirements of the Arduino often present a challenge to most new users. For example, when using Arduino, even a punctuation mistake will prompt an error message, and the code will not compile (Fig. 7.4).

Serial monitor and serial plotter: The option to open the Arduino serial monitor is available under the Tools menu item and then clicking on the magnifying glass icon; or it can be selected from the upper right side of the IDE. The serial monitor is the interface between the computer and the Arduino board, and it is used to communicate with the Arduino board using the computer. This has great utility in debugging and real-time monitoring. The serial class is needed to use the monitor. A test section from the code downloaded from https://www.circuito.io/ helps to use the serial class to test each component as demonstrated in Fig. 7.5.

The Arduino serial plotter is another component of the Arduino IDE, which allows a user to generate a real-time graph of serial data. The serial plotter makes it much easier for users to analyze the data through a visual display. With this, the user can create graphs and negative value graphs and also conduct waveform analysis, as shown in Fig. 7.6.

Fig. 7.4 Arduino code example.

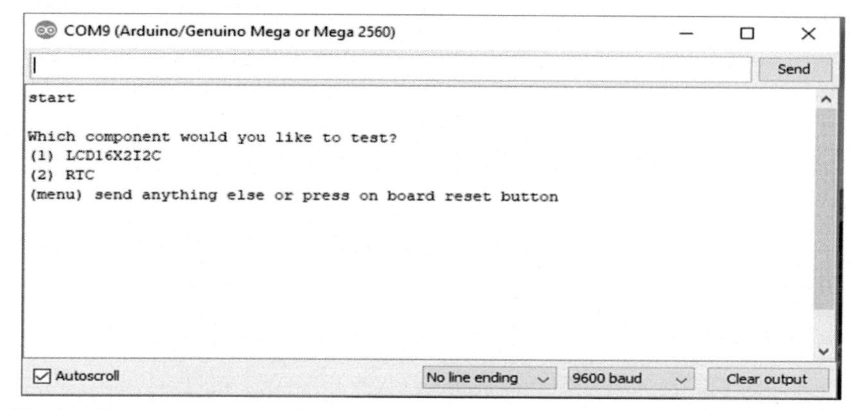

Fig. 7.5 Test section.

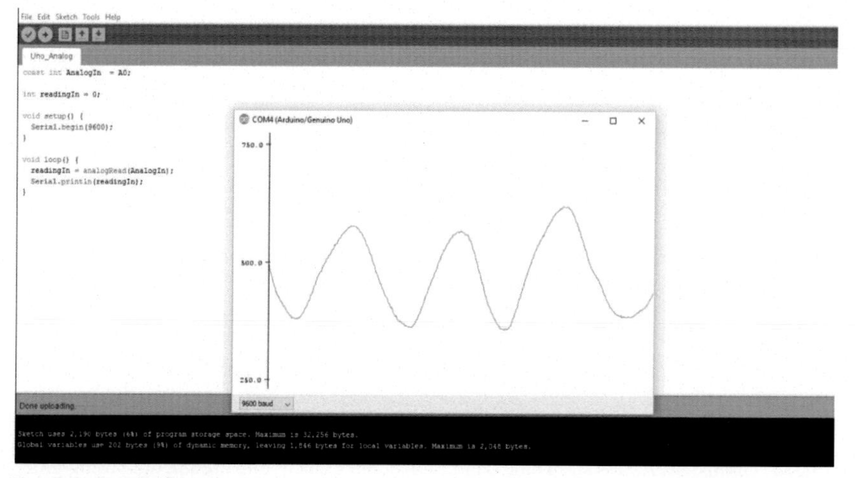

Fig. 7.6 Serial plotter.

7.3 Results and discussions

The intelligent walking stick, built with the highest degree of precision, assists the blind individuals to move from one location to another without seeking assistance from others. As testified by the 45 users during the test phase, the accuracy of the system is 97%. This could also be seen as a crude manner to give a feeling of vision to the blind. This stick decreases the reliance on other family members, friends, and guide dogs by visually impaired people while wandering around. The suggested combination of different work units allows the novel system to track the user's position in real time and offers dual feedback that makes navigation safer and more secure. The intelligent stick detects objects or barriers in front of users and, correspondingly, alerts the user, rather than vibration in the form of speech messages.

7.4 Conclusion

This paper presents a prototype for a smart assistive walking stick to guide blind individuals. Its objective is to aid in making the daily experiences of blind users less difficult as they navigate independently. This novel technology works to assist the visually impaired population of the world. The proposed system makes it simpler for the target population to walk to wherever they want without any assistance from caregivers. It is used to assist blind individuals with disabilities to promote motion and enhance safety. The aggregate performance of the system is 97% as testified by all the employed users.

References

[1] A.S. Romadhon, A.K. Husein, Smart stick for the blind using Arduino, J. Phys. Conf. Ser. 1569 (3) (2020, July) 032088. IOP Publishing.
[2] J. Poornima, J. Vishnupriyan, G.K. Vijayadhasan, M. Ettappan, Voice assisted smart vision stick for visually impaired, Int. J. Control Autom. 13 (2) (2020) 512–519.
[3] D. Chiranjevulu, D. Sanjula, K.P. Kumar, U.B. Murali, S. Santosh, Intelligent walking stick for blind people using Arduino, Int. J. Eng. Res. Appl. 10 (2020) 42–45.
[4] R.P. Soundarya, B. Reddy, A. Shankar, Smart blind stick using Arduino, Int. J. Adv. Res. Comput. Sci. 11 (2020).
[5] M. Hariharan, A.S. Raj, "Dark Light" for blind/visually impaired people using global positioning system and Arduino, Int. J. Creat. Res. Thoughts 8 (2020) 549–554.
[6] A. Pathak, M. Adil, T. Rafa, J. Ferdoush, A. Mahmud, An IoT based voice controlled blind stick to guide blind people, Int. J. Eng. Invent. 9 (2020) 9–14.
[7] J. Rwakarambi, The smart stick for the impared people, MUST J. Res. Dev. 1 (3) (2020) 10.
[8] M. Bansal, S. Malik, M. Kumar, N. Meena, Arduino based smart walking cane for visually impaired people, in: 2020 Fourth International Conference on Inventive Systems and Control (ICISC), IEEE, 2020, January, pp. 462–465.
[9] S. Budilaksono, B. Bertino, I.G.A. Suwartane, A. Rosadi, M.A. Suwarno, S.W. Purtiningrum, A.A. Riyadi, Designing an ultrasonic sensor stick prototype for blind people, J. Phys. Conf. Ser. 1471 (1) (2020, February) 012020.
[10] R. Goel, A. Jain, K. Verma, M. Bhushan, A. Kumar, A. Negi, Mushrooming trends and technologies to aid visually impaired people, in: 2020 International Conference on Emerging Trends in Information Technology and Engineering (ic-ETITE), IEEE, 2020, February, pp. 1–5.

[11] V. Kunta, C. Tuniki, U. Sairam, Multi-functional blind stick for visually impaired people, in: 2020 5th International Conference on Communication and Electronics Systems (ICCES), IEEE, 2020, June, pp. 895–899.
[12] I. Taraniya, P.V. Bhaskar Reddy, V. Chaithra, Y. Divyasri, N.L. Raviteja, Smart blind stick pro, Int. J. Adv. Res. Comput. Sci. 11 (2020).
[13] S. Karkuzhali, A. Mishra, M.S. Ajay, Design of neurocane for blind community using IoT device, in: 2020 International Conference on Computer Communication and Informatics (ICCCI), IEEE, 2020, January, pp. 1–7.

CHAPTER EIGHT

Mobile application development for hearing assistance

Ramiz Musallam Salama[a], John Bush Idoko[b], Sunsley Tanaka Halimani[c], Kevin Meck[c], and Dilber Uzun Ozsahin[c,d,e]

[a]Department of Computer Engineering, Near East University, Nicosia, Turkish Republic of Northern Cyprus, Turkey
[b]Applied Artificial Intelligence Research Centre, Department of Computer Engineering, Near East University, Nicosia, Turkish Republic of Northern Cyprus, Turkey
[c]Department of Biomedical Engineering, Near East University, Nicosia, Turkish Republic of Northern Cyprus, Turkey
[d]DESAM Research Institute, Near East University, Nicosia, Turkish Republic of Northern Cyprus, Turkey
[e]Medical Diagnostic Imaging Department, College of Health Sciences, University of Sharjah, Sharjah, United Arab Emirates

8.1 Introduction

Android is a Linux-based software that was designed mainly for touch screen hardware such as tablets and mobile phones. Android has evolved from being used in older devices and has become one of the most popular operating systems in mobile devices. Loss of hearing has affected millions of people from across the globe with the greater part of this populace living in developing countries where health services are not as efficient or readily available [1, 2]. This can cause significant challenges for those who are hearing impaired and without public health interventions, such as therapy, awareness, and rehabilitation. The purpose of the application is to aid deaf users in having a normal life. Basically, this app is for helping deaf people to communicate easily and to provide them with basic and necessary assistance [3–5]. The first section provides the ability to save numbers for different services and the ability to call these numbers in the event of an emergency. The second section is about helping the users to make phone calls, and this happens by converting their speech to text, allowing them to read audible words. This function also allows for the conversion of text to voice, which is important because not all services provide communication through text and calls.

Section one: describe the abilities and disabilities of deaf people and how the application will help them.

Modern Practical Healthcare Issues in Biomedical Instrumentation
https://doi.org/10.1016/B978-0-323-85413-9.00008-6

Section two: describe how the application works by illustrating the user interface elements and the backend part and how the functions of java work to make the app achieve the goals.

8.1.1 Capabilities of deaf people

Sign language is the means of communication adopted by many deaf people with roughly 10% of them having sign language as their first language [6–9]. Others, however, prefer to use verbal speech and lipreading because they would have lost their ability to hear at a later stage in their lives after having learned a certain language. Today, deaf people may receive medical interventions, such as cochlear implants and speech therapy, to aid in their hearing and speech. Lipreading is used as a means of communication by those who suffer from hearing loss. Unfortunately, it is difficult to learn and get used to, and much of the information cannot be grasped. Deaf people often have to rely on context and word guessing and may end up distorting what is being said. It is difficult to decode what has been said as there are many similar mouth movements when pronouncing words or letters. It is said that the best lipreaders can only tell approximately 30% of what has been said. The percentage may be a little higher if the deaf person has a close relationship with the speaker. Another disadvantage of lipreading is the amount of concentration required in addition to not being able to read multiple lips at a time. The ability of deaf people to speak depends on whether the deafness came before or after the individual learned languages prior to hearing loss. Speech can also be improved after becoming deaf with the help of speech-language pathologists. Being deaf does not necessarily translate to being mute, but verbal speech may be a challenge because of the inability to control volume and pitch when speaking. They often make sounds to emphasize phrases, expressions, or emotions. These sounds can often be understood by those with whom they regularly communicate.

Although English is the native language (mother language) for some deaf people, they can barely understand it when under pressure. They even struggle to read written English without having advanced in the language. Interpreters may be required to provide translations for the deaf person to understand what is being communicated, especially if sign language is their first language. Hearing aids are used for amplifying sound waves and may benefit some but not all deaf people because deafness comes with different defects. Some deaf people have no residual hearing, so any sound amplification has no advantage to them. Other forms of deafness come when soundwaves are distorted, making them harder to understand even when amplified.

Those who lost their ability to hear after learning to read will be able to read the text converted from speech. But those who were born deaf can read through special learning and practice. The quality of their reading may not be as advanced, but it can still aid in their use of the device and in having more independence. The application helps deaf people to be able to communicate with others by talking and primarily when phone communication is involved. This was achieved by converting the normal human voice to text so that deaf users can read it. The application also helps the deaf with texting to speech so that hearing people could understand the received communication through the phone. The application also provides the ability to save the emergency numbers in the phone. Such numbers are for the police and ambulance emergency phone lines as well as some special services phone numbers. The application also provides deaf users with the ability to make calls directly to these numbers.

8.2 Processes of creating the application

8.2.1 The UI (user interface)

It is essential that the UI has a beautiful background in order to draw the users' attention. In order to achieve this, knowledge on XML (extensible markup language) and Android View components is necessary.

XML (extensible markup language): In order to design a screen, one must be familiar with the Android Studio and the various View Components it provides.

View: View is a superclass among all the UI components and is used to design the layout of Android apps. It can be imagined as a rectangular box where elements are added.

Some of the view components included are:

ProgressBar: displays any progress or status, like an app opening loading page

EditText: for the users who want to make any inputs

TextView: used for adding text within the app

Button: initiate an action and are used to trigger some actions on the click of the button

ImageView: for any image additions that may be needed

ImageButton: can make an image clickable

CheckBox: can pick options out of the various available options.

ViewGroup: a group of Views, with the highest level being a parent and the ones under called children

These components work together to generate Fig. 8.1.

Fig. 8.1 shows the splash opening screen that contains the logo of the application with a response time of approximately 1 second, and it disappears soon afterwards.

RecyclerView: In the Android version of the proposed system, this is the view that appears when an emergency button is pressed. Here are some regulations on this feature:

- **Adapter:** This is RecyclerView.Adapter subclass that provides views representing items in a given dataset.
- **Position:** This represents the location of an item of data in a given adapter.
- **Index:** This represents the position or index of any attached child view utilized in a call to getChildAt(int).
- **Binding:** This is the method of composing a child view to show the data corresponding to the location within the adapter.
- **Recycle (view):** For later reuse, this is a view previously used to display data for a particular adapter role that can be put in a cache to show the same form of data in the future. By missing the initial layout inflation or building, this can increase efficiency dramatically.
- **Scrap (view):** A child view that, during the layout, has entered a temporary detached state. Without being completely disconnected from the parent RecyclerView, scrap views can be reused, either unmodified if no rebinding is needed or adjusted by the adapter, if the view was deemed dirty.

Text-to-speech conversion layout consists of:

- **Edit Text:** The user will use it to enter text, which will be converted to speech.
- **Text View:** This shows the text "Pitch" on the screen to label the function of the button below it.
- **Seekbar:** This is the button which controls the frequency of the voice, and it is a sliding button.
- **Text View:** This shows the text "Speed" on the screen to label the function of the button below it.
- **Seekbar:** This is the button which will control the speed of the voice, and it is a sliding button.

Fig. 8.1 The screens of the application.

- **Button:** It is a normal button, and when clicked, it runs the text in the edit text view while being converted to speech.
- **Text-to-Speech function:** This is a function provided by Google to convert the text to speech.
- **Speech-to-Text function:** It is the function responsible for converting the speech to text.
- **Pitch control:** It is the function that changes the frequency of the voice. If the value of pitch is low, the voice will sound like a male voice. If the value of pitch is high, the voice will sound like a female voice.
- **Speed control:** This function changes the speed of speech.
- **Set text:** This function sets the words from the speech to the screen.

8.2.2 Speech recognition

Various complex steps are taken by the computer in order to convert speech into a computer command or into on-screen text. Air vibrations are created when one speaks. To translate the vibrations, which are in the form of an analog wave into digital data that the computer understands, the analog-to-digital converter (ADC) is used [7]. The ADC performs this function by digitizing, or sampling, the sound by taking exact measurements of the wave at specific frequent intervals. To get rid of the unwanted sound that is called the noise, the system filters the digitized data, or in other instances, the noise is split into frequencies of different bands (the wavelength of a specific sound is called the frequency, and this is perceived as different pitches to the human ear). The sound is then adjusted to a constant volume and aligned; or in other words, it is normalized. The sound must also be tuned to match the speed of the already stored template sound samples because humans do not always speak with the same speed.

The signal is then split into sections that are a few hundredths of a second and even as short as a few thousandths when it comes to plosive consonant sounds; these are consonant stops that are created by blocking airflow in the vocal tract, like a "t" or a "p." The program proceeds to match the segments to known phonemes in the relevant language. A phoneme is the smallest unit of a language used to establish distinctions between words [10]. The English language has approximately 40 phonemes (linguists debate over the exact number), while other languages can have either fewer or more phonemes. Although seemingly easy, the most difficult step to accomplish follows. This step is also the primary focus of most speech recognition research. The phonemes in the context of the other phonemes around them are assessed by the program. The contextual phoneme plot through a complex statistical model is run and compared to a huge library of known words, phrases, and sentences. The program proceeds to determine what the user was likely saying, and it will output it either as a computer command or text.

Complex and powerful statistical modeling systems are utilized in today's speech recognition systems. To determine the most likely outcome, the statistical modeling systems use probability and mathematical functions. Neural networks and the Hidden Markov Model are the dominant models in the field today according to leading speech expert John Garofolo of the National Institute of Standards and Technology. Complex mathematical functions are a feature of these methods, and they fundamentally take the data that is already known to the system and then try to factor out the data that is hidden from it.

8.2.3 Working principle of text-to-speech

Speech synthesis is the production of human speech through artificial means by the use of a speech computer/speech synthesizer, either integrated in software or within the device hardware. To convert a text into speech, a text-to-speech (TTS) is used.

Artificial speech is made through linking together small parts of prerecorded speech, which is kept in a database. Stored speech units may vary in size with those storing phones/diphones having the greatest range of output, but clarity is a tradeoff. Storing words, phrases, or complete sentences gives an output with better quality and clarity. The synthesizer can also use human voice features to create a more realistic or synthetic voice output. The ability to mimic the human voice determines the overall quality of the speech synthesizer. The quality of a speech synthesizer is judged by its similarity to the human voice and by its ability to be understood clearly.

A text-to-speech system consists of a front and back-end (synthesizer). The front-end is responsible for text normalization preprocessing, where it converts symbols and abbreviations into written words. This can also be referred to as tokenization. After tokenization comes text-to-phoneme/grapheme-to-phoneme conversion, in which each word is assigned a phonetic transcription followed by the separation and marking of the text into prosodic units. The phonetic transcriptions and prosody information will then become the components of the output of the front-end that is the symbolic linguistic representation. The back-end transforms the output of the front-end into sound.

Reading words has some complications. These complications come when words have more than one meaning, and without the knowledge of these meanings, it makes it harder to understand when reading. The first stage in synthesizing speech, preprocessing normalization, aims to minimize any ambiguities by determining which meaning may be the most suitable through proofreading the text to avoid any reading errors. Figures, times, symbols, dates, abbreviations, etc. have to be read out as words, and this may distort their meaning. Computers cannot differentiate numbers from a year from numbers of quantity although they vary in pronunciation. To help them, computers use statistical probability methods in which they may guess the meaning based on context. For example, if users put the word "date" in the same sentence with a number, it may assume the number was referring to the date or year. It uses the same concept for words that can have different meanings such as the word "letter." The computer would then have to try to discern the meaning of the word using the context of the sentence.

8.3 Conclusions

Deaf people can face issues in communication especially if they live alone. They can acquire some skills, such as learning how to read, interpret body language, and lipread. The application provides the features that could help deaf people in communicating more easily.

The application consists of two parts: the user interface and the backend; each interface page relates to a code page that controls this interface. The application has six interface pages with six code pages.

The aim of this application is to help those who are deaf and affected by hearing loss to communicate in an easy way and give them the assistance that they need. The application helps them to communicate with people that may not understand sign language and other hand/body gestures and helps them to communicate with people who are in front of them (through phone calls) where body language and lipreading would not be helpful. The proposed application also provides the target population with the ability to save and call numbers to get help in the event of emergencies or to get some services. This assistance is possible by the help of the text-to-speech function and the speech-to-text function. Here, the voice of a hearing individual is converted into text that a deaf user can read, and the deaf user would write what they want to communicate for this text to be converted into voice for the hearing individual. And this process makes the communication to easier, especially in day-to-day challenges faced by members of the deaf community.

References

[1] C. Ceccarini, C. Prandi, Tourism for all: a mobile application to assist visually impaired users in enjoying tourist services, in: 2019 16th IEEE Annual Consumer Communications & Networking Conference (CCNC), IEEE, 2019, January, pp. 1–6.
[2] M. Debevc, J. Weiss, A. Šorgo, I. Kožuh, Solfeggio learning and the influence of a mobile application based on visual, auditory and tactile modalities, Br. J. Educ. Technol. 51 (1) (2020) 177–193.
[3] D.H. Hanes, U.S. Patent No. 10,580,266, U.S. Patent and Trademark Office, Washington, DC, 2020.
[4] M. Hossain, M. Hasan, M. Riton, A. Islam, Usability Evaluation of Bangladesh Government Mobile Application, 2019.
[5] V.V. Kumaran, S.S. Nathan, A. Hussain, N. Hashim, Mobile banking usability evaluation among deaf: a review on financial technology and digital economy prospects, Int. J. Interact. Mob. Technol. 13 (2019) 24–33.
[6] O. Lozynska, V. Savchuk, V. Pasichnyk, Individual sign translator component of tourist information system, in: International Conference on Computer Science and Information Technology, Springer, Cham, 2019, September, pp. 593–601.
[7] L. Olga, S. Valeriia, P. Volodymyr, The sign translator information system for tourist, in: 2019 IEEE 14th International Conference on Computer Sciences and Information Technologies (CSIT), vol. 3, IEEE, 2019, September, pp. 162–165.
[8] R.L. Romero, F. Kates, M. Hart, A. Ojeda, I. Meirom, S. Hardy, Modifying the Mobile App Rating Scale with a content expert: evaluation study of deaf and hard-of-hearing apps, JMIR mHealth and uHealth 7 (10) (2019).
[9] R.L. Romero, F. Kates, M. Hart, A. Ojeda, I. Meirom, S. Hardy, Quality of deaf and hard-of-hearing mobile apps: evaluation using the Mobile App Rating Scale (MARS) with additional criteria from a content expert, JMIR mHealth and uHealth 7 (10) (2019), e14198.
[10] G. Wojtanowski, C. Gilmore, B. Seravalli, K. Fargas, C. Vogler, R. Kushalnagar, "Alexa, Can You See Me?" Making individual personal assistants for the home accessible to deaf consumers, J. Technol. Persons Disabil. (2020) 130–148.

CHAPTER NINE

Development of medical dispatcher: A robot that delivers medicine

Dilber Uzun Ozsahin[a,b,c], John Bush Idoko[d], Mandy Sizalobuhle Mpofu[b], Ismail Ramadan Swalehe[b], and Ilker Ozsahin[a,b,e]

[a]DESAM Research Institute, Near East University, Nicosia, Turkish Republic of Northern Cyprus, Turkey
[b]Department of Biomedical Engineering, Near East University, Nicosia, Turkish Republic of Northern Cyprus, Turkey
[c]Medical Diagnostic Imaging Department, College of Health Sciences, University of Sharjah, Sharjah, United Arab Emirates
[d]Applied Artificial Intelligence Research Centre, Department of Computer Engineering, Near East University, Nicosia, Turkish Republic of Northern Cyprus, Turkey
[e]Brain Health Imaging Institute, Department of Radiology, Weill Cornell Medicine, New York, NY, United States

9.1 Introduction

A medical dispatcher robot is a special type of robot used to deliver medication to patients. The medical dispatcher, controlled by a user behind the computer system, delivers medicine to different patients in the hospital [1]. The name "medical dispatcher" was derived from the need to have a form of delivery service for those patients who require their medication to be administered to them. The outcome of this research will assist medical facilities that are lacking nurses by incorporating technologies to dispatch medications to patients. This will help medical settings to reduce staff numbers, which will in turn save managerial cost [2]. In considering what a successful medical dispatcher would be, it should be able to travel at least in a 120 m^2 region, deliver the medication uncontaminated, avoid all sorts of collisions, transport the medication fast and efficiently, suit the user's needs with adjustable cameras, communicate with the patients, be user friendly, and need very little calibration over long periods of time [1].

Modern Practical Healthcare Issues in Biomedical Instrumentation
https://doi.org/10.1016/B978-0-323-85413-9.00002-5

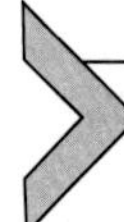

9.2 Components of the medical dispatcher

9.2.1 Arduino Uno

The Arduino Uno is a microcontroller board that uses the ATmega328 datasheet. It has 14 digital input/output pins, 6 analog inputs, a power jack, a reset button, and a USB connection [3]. These are used to control some of the parts of the medical dispatcher.

9.2.2 Wheels

The wheels are controlled using the motherboard. The front wheels have the capacity to rotate up to 270 degrees. The back wheels are used for support. For continual efficiency, the wheels should be replaced after long periods of use; otherwise, they should be cleaned and maintained.

9.2.3 Motherboard

A motherboard is a printed circuit board that uses the key components of a computer. This is where the heavy programming of the medical dispatcher takes place. The utilization of all the ports and keyboard functions plays a great role in controlling the device. The controls are sent and received by the motherboard. All connections are linked to the motherboard.

9.2.4 Airtight carrier

This is a container used to store the medication delivered to patients. The container has to be airtight for the medication to remain uncontaminated.

9.2.5 Camera

Two cameras are located on the front and back regions of the medical dispatcher robot. The camera is adjustable so as to control what the user views.

9.2.6 LED lamps

Light-emitting diode (LED) lamps are used to produce light. They are not harmful, and they are cheap to replace. They are meant to indicate that the medical dispatcher is moving and can be activated to generate a flash upon reaching the user. The LED lamps can also be used to light up the path that the robot is using if the environment is dark.

9.3 Construction methodology

This section demonstrates how all of the components are interconnected to achieve the set objective. The proposed gadget is controlled by connecting it to a computer system using Bluetooth, an internet connection, and an infrared detector. The wheels are connected to a motor and a battery and are controlled by the Arduino for movement. The cameras are connected to the front and back sides of the robot to monitor the entire movement of the robot, and the user will be able to see the patient on the monitoring system when the robot arrives at the destination. A speaker and microphones are also connected to the main board to enable the user to communicate with patients at delivery points. Lamps are placed on the device for different functions: two lamps at the left and right sides of the robot to indicate that it is moving, a third lamp connected to alert the patient to receive their medication, and a fourth lamp placed at the front to brighten the path at dark times. Lights and other components, such as the infrared obstacle sensor, are programmed by the Arduino Uno to activate an action. The infrared sensor is placed at the front of the medical dispatcher to detect obstacles. Batteries are connected to the respective power sources to power the entire system. Figs. 9.1 and 9.2 represent the block diagrams for the road and air dispatchers, respectively.

One advantage of the proposed device is that it creates a new pathway for communication between the patient and the medical expert, thus saving the user's time in avoiding time-consuming walks from one patient to another as the medical expert can talk to more than one patient at once while seated in one location. Secondly, pharmacists will have enough time to serve other patients because less time will be wasted going from one patient's room to another in dispensing and delivering medication; the pharmacist can easily pack the medicines into different drawers of the carrier, and these medicines would be controlled for a safe delivery. Thirdly, quarantine centers could use the medical dispatcher to administer medicine while preventing the spread of transmittable diseases. The fourth advantage is that it will reduce the cost of employing staff who would usually perform these services. Finally, the medical dispatcher is a multitasking robotic that can do the delivery as well as convey messages. The robot has a few limitations, which include (1) medication cannot be administered from the medical dispatcher to the most critical patients (critical cases in the intensive care unit); (2) it is not useful to some disabled patients, such as those experiencing dwarfism, deafness, and

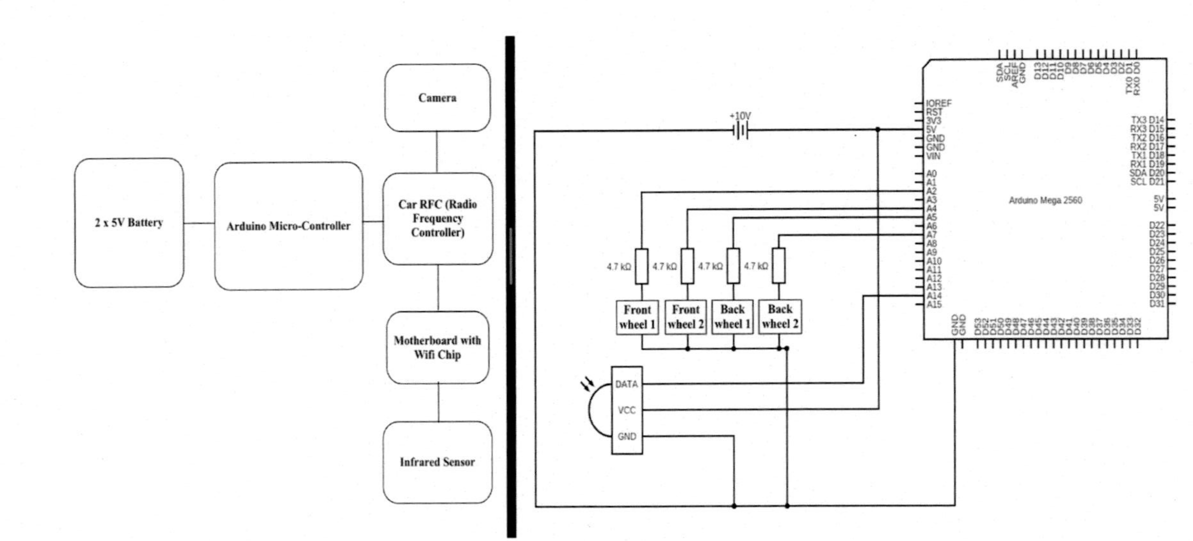

Fig. 9.1 Road dispatcher circuit diagram.

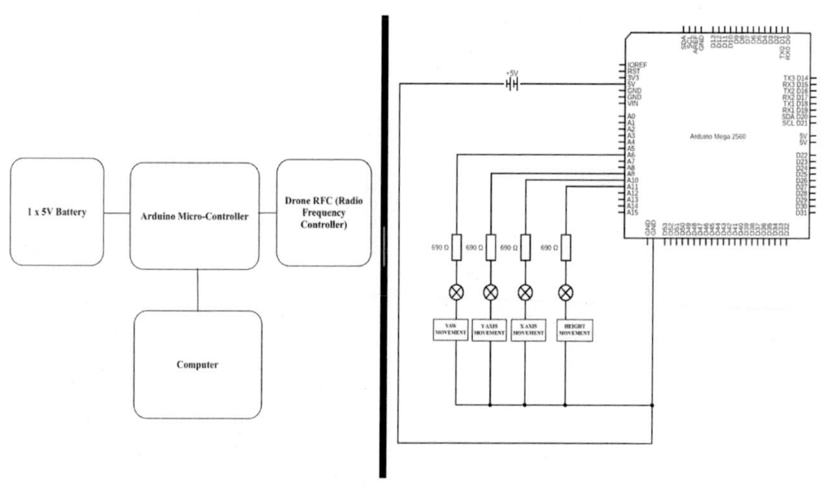

Fig. 9.2 Air dispatcher circuit diagram.

blindness; (3) the robot is not helpful for illiterate patients because they may not understand the instructions given by the robot; and (4) it can be easily destroyed by other people during transit because most people may assume it to be a toy.

9.4 Results and discussion

This section illustrates the results obtained from the performed experiments. Multiple tests were performed on the medical dispatcher. For the device's capacity for movement, the forward and backward, left and right movements were successfully tested with satisfied outcomes. At the test phase, we were able to identify some mechanical and software errors that were technically and appropriately corrected before the first medication delivery was made. The road transporter model was able to achieve the communication and medication delivery, providing safe transportation of medicine from one point to the another. The interface of the road dispatcher was easier to achieve as the coding section is less complex. The employed system controller/medical expert and the patients on average rated the performance of the device at 90%.

Unlike the road medical dispatcher, the medical drone has some complexity in the coding and movement section; the debugging has multiple problems (consequently corrected) in matching communication ports to

Table 9.1 Tested features.

	Arduino Uno	Arduino Uno Software	Matlab Software	Left and Right Movement	Forward and Backward Movement	Upward and Downward Movement	Temperature Sensor	Camera
Medical Dispatch: Road	✓	✓		✓	✓		✓	✓
Medical Dispatch: Air	✓	✓	✓	✓	✓	✓		

both MATLAB and Arduino. Initial tests were performed to get the required weight that would be necessary for the drone system as weight can affect its trajectory due to the motors. The advantage is that the bigger the motors, the bigger the weight that can be carried. The average rating by the employed system controller/medical expert and patients is approximately 87%–94%. Table 9.1 shows the tested items ensuring the mobility of the medical dispatcher.

9.5 Conclusions

As seen in the results, both the models achieved the main goal in delivering medicine to patients. The road dispatcher uses the camera and infrared sensor for its movement. The controllers of robot were able to track the movement and location of the device at any given point. The movement was actually possible without a camera; however, no report could be made without the use of the camera. The air dispatcher is a bit more complex than the road dispatcher. It requires the user to know the coordinates in which the device flies. The addition of other components to the device affects its trajectory; however, the addition of larger motors can enable the drone to carry larger loads. The testimonies from the users and patients show that devices are ready for distribution and use. Future exploration of this study would focus on advancing these dispatchers into a smart robot that learns and communicates with patients, thus reducing the need of communication between the users and the patients.

References

[1] M. Woollard, G. Jones, Post-dispatch and pre-arrival instructions—Wales' experience, in: J.J. Clawson, K.B. Dernocoeur, B. Rose (Eds.), Principles of emergency medical dispatch: 30 years of protocols (1979-2009), fourth ed., National Academy of Emergency Medical Dispatch, Salt Lake City, UT, 2008, pp. 8.8–8.9.
[2] B. Toon, Life-impacting via telephone instructions, in: J.J. Clawson, K.B. Dernocoeur, B. Rose (Eds.), Principles of Emergency Medical Dispatch: 30 years of protocols (1979-2009), fourth ed., National Academy of Emergency Medical Dispatch, Salt Lake City, UT, 2008, pp. 1.10–1.11.
[3] N.S. Kumar, B. Vuayalakshmi, R.J. Prarthana, A. Shankar, IOT based smart garbage alert system using Arduino UNO, in: 2016 IEEE Region 10 Conference (TENCON), Singapore, 2016, pp. 1028–1034, https://doi.org/10.1109/TENCON.2016.7848162.

CHAPTER TEN

Design and implementation of wireless helmet and mechanical wheelchair

Dilber Uzun Ozsahin[a,b,c], John Bush Idoko[d], Mohamad Hejazi[b], Rayan Allaia[b], Mennatullah Ahmed[b], Zuhdi Badawi[b], and Ilker Ozsahin[a,b,e]

[a]DESAM Research Institute, Near East University, Nicosia, Turkish Republic of Northern Cyprus, Turkey
[b]Department of Biomedical Engineering, Near East University, Nicosia, Turkish Republic of Northern Cyprus, Turkey
[c]Medical Diagnostic Imaging Department, College of Health Sciences, University of Sharjah, Sharjah, United Arab Emirates
[d]Applied Artificial Intelligence Research Centre, Department of Computer Engineering, Near East University, Nicosia, Turkish Republic of Northern Cyprus, Turkey
[e]Brain Health Imaging Institute, Department of Radiology, Weill Cornell Medicine, New York, NY, United States

10.1 Introduction

The term disability entails a very wide spectrum of conditions that signify difficulty in certain activities and/or interaction with the surrounding environment. This spectrum is categorized and defined by the nature of the disability or impairment; it could be cognitive, developmental, intellectual, mental, physical, or even sensory. These conditions are more common than they appear.

According to the World Health Organization, more than a billion people are approximated to have some kind of disability; this makes up about 15% of the worlds' population as of 2010. Based on previous statistics, this number appears to be increasing, indicating that the number of people dealing with some form of disability is increasing. This could be explained by the prevalence of an aging population as well as the rise in the occurrence of chronic health conditions related to disabilities; for instance, diabetes, cardiovascular disease, and mental illness are all chronic diseases that can eventually escalate to cause disabilities. In terms of the geographical distribution of disabilities, the propagations will be dependent on multiple factors including, but not limited to, trends in health conditions, trends in environmental factors, diet, substance abuse, and even road crashes and conflict, be it war conflicts or

Modern Practical Healthcare Issues in Biomedical Instrumentation
https://doi.org/10.1016/B978-0-323-85413-9.00014-1

criminal activities. From this, it is glaring that the prevalence of disabilities is more persistent in lower income countries. People living with these impairments struggle to carry out day-to-day tasks, which makes their quality of life face a significant deficit. Their lives are inherently of equivalent value to any other life, so they deserve to be given equal resources and opportunities. As a society, there must be core values and morals to accommodate for the difficulties faced by those with any form of disability [1]. With the help of technological advancement in both medical and engineering disciplines, many successful tools and machinery have been developed to provide some kind of aid to lessen the effects of many types of disabilities. Hearing aids have been developed and constantly upgraded to help those with hearing impairments; prosthetic limbs have also been developed to mimic the lost function of amputees. Bionic eyes are being used to restore vision loss, and many more developments exist and continue to be discovered and advanced in the name of empathetic and useful technological design.

The proposed system is aimed to be among this list of significant advances in the field of physical impairments and disabilities. This study is intended to provide disabled individuals with an artificial transportation medium in a more independent manner than the conventional mechanical wheelchair. In this way, the user can be given the possibility of being more self-reliant, which is considered to be a highly significant change. The deep-rooted point of view offers approaches to investigate the various experiences of handicap in various settings, including small-scale and large-scale measurements [2, 3]. For example, two people living with physical debilitation will have different life experiences and thus have different views of wheelchair control. In their methodical survey of subjective investigations on maturing with an incessant or impairing condition, detailed that lingering side effects and age-related changes influenced physical capacity, personality, and self-recognition [3, 4]. Smart wheelchairs have been the subject of research since the mid-1980s and have been created on four continents [5].

We decided to focus on the category of patients with paralysis as well as amputees suffering from mobility and physical impairments, specifically from the neck down. This choice was inclined by the increase in the number of patients needing wheelchairs. There are several electric wheelchairs that are controlled by hands or feet for use by people with special needs to move forward, backward, and sideways. Because it is controlled by hands and feet, it cannot be used by people with special needs who cannot control their limbs. The present invention introduces a new electronic seat designed for people with special needs. It provides more help to people who have suffered a stroke or have paralysis but can move the head. Because this target population cannot move their hands or legs but can move their heads, this invention can expand their circle of movement, allowing them to roam in open and crowded places and to move through buildings without fear; and this can give them a sense of medical independence as they would be given the opportunity to carry out their day to day tasks with minimal help from others. This study is aimed to improve the quality of life for patients who suffer from paralysis.

There are several types of wheelchairs available on the market that can be controlled mechanically by your hands or feet. These are used by people with special needs to move forward, backward, and steer. However, these wheelchair models cannot be used by people with special needs who cannot control or have lost motor function of their limb or limbs. A wheelchair is a chair dedicated to the replacement of walking, and it comes in various shapes that can be driven by an engine or the person using it by rolling the rear wheels by hand. There are often handles behind the seat for someone else to push the chair. The wheelchair is used by people who find it difficult or impossible to walk because of illness (psychological or physical), injury, or disability [6].

Several methods have been proposed to allow people with special needs including those with quadriplegia, which is a condition that refers to paralysis in all four limbs, to control a motorized chair. The types of wheelchairs vary in terms of the functions available, the mechanism of work, and the materials used in their manufacturing. The prices of ordinary wheelchairs also differ from those of electric wheelchairs. The wheelchair is designed as an alternative to walking and to fix and correct the sitting position of the user while becoming comfortable with moving from one place to another. Here are a list of several types of wheelchairs and their differences:

Terrain wheelchairs, manual and powered: Off-road wheelchairs have developed from a standard wheelchair with larger than usual haggles to handle more difficult terrain. Many particular highlights, for example, tank tracks, hard core engines and batteries, uncompromising or fortified edges, augmented suspension, and inflatable tires have made the off-road wheelchair a preferred choice for wheelers who need to come back to nature or whose work has them out in difficult territory [7].

Manual wheelchairs: These are wheelchairs that are moved by classic muscle power or on account of transport wheelchairs that can be pushed by someone strolling behind the wheelchair. There are different kinds of manual wheelchairs that extend from ultra-lightweight to hardcore. More explicitly, manual wheelchairs fall into at least one sort or style of wheelchairs [8].

Powered wheelchairs: These are battery-controlled wheelchairs that can be constrained by the client or by a nearby specialist. These specialists prepare patients with extreme inabilities and portability misfortune to become more autonomous [9]. Progressed fueled wheelchairs are adaptable enough to be adjusted to practically any requirements with the expansion of specific parts, for example, a respirator, discourse gadget, natural control, particular controls and switches, and an incredibly inexhaustible determination of situating and body arrangement alternatives. They can be hard in front wheel drive and back haggle wheel drive. For those wheelchair clients who need not bother with a progressed fueled wheelchair framework, there are various different sorts that run from very light organizers to heavy models.

Pediatric wheelchairs: These types are dedicated and intended for children. Pediatric wheelchairs can be fitted with situating gadgets and segments like those found on grown-up wheelchairs. Most will have changes, for example, backrest stature, seat width, and armrest and leg rest tallness, that take into account development [10]. It is significant that children rapidly and effectively conform to utilizing a wheelchair and feel great cooperating with peers from a wheelchair. A cool watching seat decorated in the manner that a child likes goes far in building a child's confidence and certainty. Most makers of pediatric wheelchairs currently offer a line of fun-looking wheelchairs and embellishments.

Positioning wheelchairs: Being in the correct position while in a wheelchair can have a major effect on your wellbeing. Physiological capacities should not be restrained; they should be assisted in further development and growth. That is what positioning wheelchairs do best. Another significant purpose for changing positions while in a wheelchair is to assuage pressure on hard pieces of the body that are inclined to tissue injury. The present situating wheelchairs are exhibited by a lot of wellbeing and health advertisers. Many can change the wheelchair situation as needed. The tilt function tips the whole seating framework back without changing the seat to back point [11]. The recline function leans back the back to a near lying down position. The elevation of the legs can rest from a bowed situation to a straight-out position and any place in between in that range.

Standing wheelchairs: Every year, a large number of people get injuries in their spinal cord, and most of them use wheelchairs to be able to move [12].

Despite the quality of wheelchairs, there are defective wheelchairs that harm patients much more than they help, and in the long term, they are exposed to diseases such as hyperkalemia [13], severe osteoporosis, and increased risk of fracture and joint cramps [14]. This may limit functional tasks, and most of these diseases appear due to the position of sitting mucous over the years, and for this reason, scientists invented the standing chair to help patients. There may be some benefits in providing alternative seating positions after spinal cord injuries. This invention allowed patients to stand again and move their bodies. This chair works in a different way; when the patient stands up, the chair supports him. Standing is physiologically an excellent experience for individuals to have. Besides the physiological advantages there are other benefits bundled in these standing wheelchairs: the capacities to confront society on an eye-to-eye level or the capacity to reach places not ordinarily reachable from a standard wheelchair height. Users can't simply thud down in a standing wheelchair. There are various cushions and ties that must be situated accurately so as to encourage standing. Like any unpredictable innovation, things could become ineffectual or costly in the long run as these units can become costly to fix.

Plane wheelchairs: This is another special item, fundamentally utilized via aircrafts to enable transport of individuals who are disabled. They will in, general, be smaller and lighter than customary manual wheelchairs as they are intended to go between the aisles of seats on a plane, and they feature clasps with the goal that they can be locked into place during a flight, much the same as the travelers' seats. It is completely conceivable, obviously, to utilize them every day, even in nontraveling contexts.

Reclining wheelchairs: This is an amazing niche design, presenting all the benefits of a recliner, rolled into a wheelchair. These designs come with a pillow to offer more comfort when resting in a reclined position. With this functional comfort, it is feasible to nap in chairs of this kind.

10.1.1 Aim of the study

This study presents a wireless helmet and wheelchair access control system. It focuses on building a smart wheelchair controlled by head gestures to move it to an area or location as directed. The system consists of two main parts: the transmitter and the receiver. The sensor in the transmitter is designed to send signals to the receiver. The system also consists of a special wearable helmet worn on the head. This wheelchair can move left and right as directed by the user. This study introduces a new electronic wheelchair designed for people with special needs. It provides more help for people with stroke or paralysis. As these people cannot move their hands or legs but can move their heads, we designed a system controllable by their head gestures. Accordingly, this system can aid these people in moving without the help of others. The dependency on a caregiver or a family member has been proven to increase the chances of developing some sort of mental strain and, eventually, disorders, such as depression and anxiety.

10.1.2 Background of the study

This invention relates to an electric wheelchair that is controlled by the movement of the head, as incorporated into the system. The wheelchair has an electric wheelchair structure and helmet-shaped head signals. The helmet is portable with electrodes, acceleration sensors, and a signal module and unit. An nRF transmission module; a wireless receiver module; a microcontroller module, operating with a motor in the wheelchair configured to communicate with the user's head movement; and a communication interface device connect to the wheelchair control unit for signal processing.

10.2 Materials and methods

10.2.1 Components

The system requires a total of 12 components to operate as needed. Each of these components provides their respective functions that eventually communicate with the overall system to yield the end product. Each one of these components is crucial for the system to function the way it is intended. In Section 10.2.1.1, the system will be dissected into its individual elements, where the working principle of each of the elements and how they function to support the overall end goal of the study will be elaborated. For the wireless helmet and mechanical wheelchair system, we utilized 12 main electronic components.

10.2.1.1 Microcontroller board (Arduino Uno)

As seen in any electrical project, a central processing unit is required to carry out the set of functions required. Generally, microprocessors are favorably used in most electrical and digital devices because of their availability, associated convenience, and flexibility; as there are many types that offer different features and thus carry out a very wide range of functions. These traits make microprocessor and microcontroller boards easily applicable in most, if not all, digital devices [15]. There are quite a lot of types of microprocessors on the market ranging in complexities and features. For the purpose of this study, we decided to implement the Arduino Uno because it is widely available, user friendly and easy to use, especially for educational purposes and projects like this one, due to its distinguishable simplicity and familiarity. Also, it is an open-source platform, meaning that anyone could use it, again emphasizing its availability and accessibility. Arduino is a development board with a software environment for the Atmel AVR microcontroller [16]. This panel is easily programmed. We only needed a USB cable to download the Arduino IDE software environment. The Arduino is a microcontroller board composed of an integrated circuit that carries out the functions of a computer processing unit. This specific model, Arduino Uno, has an onboard voltage regulator, type B USB connector, reset button, power jack, in-circuit serial programmer (ICSP), 3.3 and 5 V output power ports, 2 ground voltage ports, 6 analog input ports numbered from 0 to 5, and 14 digital ports numbered from 0 to 13. In total the Arduino Uno has 33 ports, including both analog ports and digital ports. This model can be connected to a computer using either the USB connector port or the charger port. The charger port receives a voltage of 7–12 V to be adjusted internally to 5 V. The Arduino Uno has six digital ports that support pulse width modulation. There are six ports in the center of the board. These ports are used to program Arduino from an external programmer called an ICSP header. The reset button on the panel functions to restart the program loaded on Arduino. We have found that powering the Arduino from the power jack, as opposed to the charger port, will minimize the amount of noise in the signal and therefore lessen the filtering required. In our study, we have implemented two Arduino boards for each of our two circuits: a receiver circuit and a transmitter circuit, both of which have been programmed appropriately to suit the function of the circuit. For the transmitter circuit, the Arduino board is connected to the accelerometer from the analog input ports/pins and then to the nRF transmitter from the output pins [17].

The Arduino platform has become well acquainted with humans in electronics. Unlike most preceding programmable circuit boards [18], the Arduino does not have a separate piece of hardware to load new code onto the board; a USB cable can be used to upload, and the software of the Arduino uses a simplified version of C++, making it easier to program; it offers an easier environment that ignores the features of the microcontroller into a more reachable package. An Arduino board can be categorized into two components: the hardware part and the software part.

10.2.1.2 MEMs accelerometer: ADXL355 model

The accelerometer is the second component in our design, and it is the most fundamental component. This device measures and calculates acceleration, as suggested by its name. The microelectromechanical system (MEMS) accelerometer is a component used in electric circuits to provide structural measurement of constant acceleration, i.e., a motion sensor. MEMS are used to describe a wide range of sensors that are produced or manufactured using microelectronic fabrication techniques [19].

The sensor component is composed of a series of differential capacitors with independent fixed plates, mobile plates attached to the moving mass, and some poly-silicon springs. Acceleration causes deflection in the moving mass and, in turn, initiates an unbalance in the differential capacitors as the distance between the capacitor plates are affected. In this way, the sensor output signal will be proportional to the acceleration amplitude. This output signal will be in analog form. Afterwards, the signal will be amplified by the AC amplifier and filtered based on the user's bandwidth preference using the capacitors Cx, Cy, and Cz to give the final outputs X out, Y out, and finally Z out. Even though the sensor has the ability to sense acceleration in three dimensions (x, y, and z), only two dimensions are used in this study to represent backwards and forwards directions (y-axis) as well as right or left directions (x-axis). Because of this, only outputs X out and Y out are connected to the Arduino Uno.

An accelerometer is a mass-damper-spring framework [20]. The most well-known accelerometers are either founded on piezoelectric impact or capacitance. Other detecting procedures incorporate electron burrowing, reverberation, and optical- and warm-based sensors. Piezoelectric accelerometers rely upon the pressure created by the power on the miniaturized scale precious stone structures, subsequently producing a voltage. Capacitance-based accelerometers sense the variety in capacitance because of the development of the plates. The ADXL355 consists of an inside configurable digital bandpass filter. Both the high-pass and low-pass poles of the filter are adjustable [21]. At power-up, the default prerequisites for the filters are as follows: high-pass filter (HPF) = dc (off), low-pass filter (LPF) = a thousand Hz, output statistics price = 4000 Hz. There are three units of MEMS sensing factors that follow separate differential sign trajectories to minimize noise and drift. The ADXL355, being a digital variant of a MEMS accelerometer, includes three analog-to-digital converters (ADC) to furnish high decision 20-bit digital output per axis. The analog antialiasing filter filters the X, Y, and Z axis analog outputs before entering the high-decision ADCs. A successive approximation register (SAR) ADC is used with the aid of the temperature sensor to supply 12-bit decision digital output.

10.2.1.3 Relay 5 V

One of the most important electrical elements in electronic circuits is a mechanical switch that is electrically controlled by a voltage applied to the coil inside. The microcontroller-based overcurrent hand-off has some of the same characteristics as converse unequivocal least time inverse definite minimum time lag hand-off [22]. The operation is exceptionally straightforward in that the line current is detected by the CT [23]; the detected current is at that point bolstered to the current sensor ACS-712, which in advance gives this current to Arduino in voltage frame. The current, which is bolstered to Arduino, is handled to check whether this current is more prominent than the set pick up current.

10.2.1.4 Buzzer

A buzzer or beeper is an audio signaling device that is mechanical, electromechanical or piezoelectric. An active buzzer uses an internal oscillator to generate a tone, so all that is needed is a DC voltage. To make a sound, a passive buzzer requires an AC signal. It is like an electric microphone where the data changes (Fig. 10.1).

10.2.1.5 Breadboard

A breadboard is a board used to connect electronic components, such as wires, resistors, capacitors, and coils, to conduct various experiments and projects. The panel consists of several segments, and the holes serve as connecting points where the various electronic terminals can be fixed as they are connected respectively to the surrounding holes horizontally in the two segments on the edge of the board, where the holes surrounded by green are connected straight as depicted in Fig. 10.2 [24].

Fig. 10.1 Buzzer component.

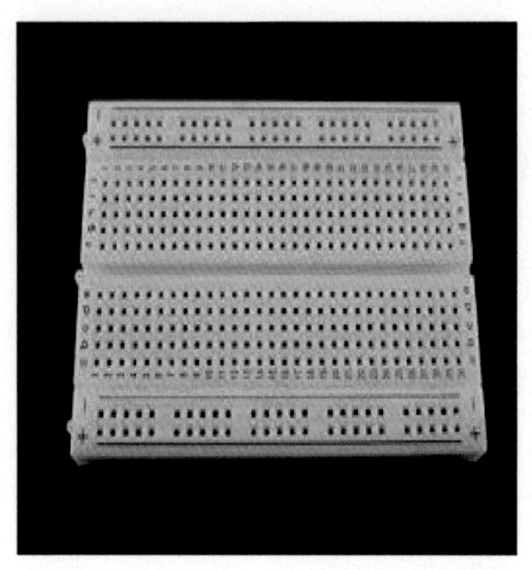

Fig. 10.2 Breadboard.

10.2.1.6 Transistor

Transistors are electronic elements made from semiconductors and used as either electrically controlled switches or as amplifiers. In a transistor, the current is controlled by a voltage or a small control current applied to the end of the control lead, and a large current is passed between the other ends of the transistor. The transistor has three leads, as seen in Fig. 10.3, one of which is the controller. Transistors are used in most electronic circuits, such as amplification circuits and vibration circuits, current source circuits, voltage regulating circuits, supply source circuits, building integrated circuits, and control circuits, especially when a small current is used to control a large current. Transistors are also used as electronic switches. It can be considered a variable resistance, and this resistance changes by changing the base current.

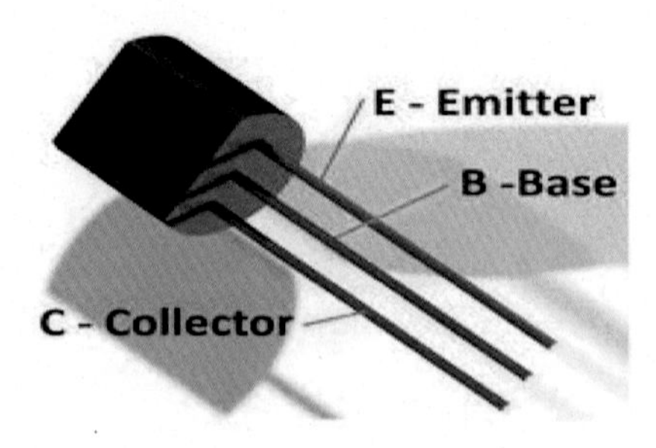

Fig. 10.3 Transistor component.

10.2.1.7 DC motor

The DC system consists of two main parts: the magnetic circuit that generates magnetic force lines and the agitator circuit in which the electric motive force is generated. The DC motor consists of the inductor and collector. The inductor is made of several electromagnetic electrodes containing central coils whose function is to generate magnetic flux. These coils are called irritation coils. The agitator consists of a magnetic core composed of electromagnetic iron sheets and has ducts in which the agitator coils are located. The collector consists of copper sheets isolated from each other and the axis of the machine, and the electrical current is taken from the coils of the agitator from the collector through two charcoal levers in contact with the collector plates. The main magnetic flux is generated as a result of the constant electric current passing in the irritation windings, where the main magnetic flush from the North Pole enters the South Pole, passing through the agitator nucleus and cutting its coils, and this flux closes its path through the fixed object through the magnetic iron core. The DC motor is used to generate continuous motion whose speed is easily adjustable, making it ideal for use in applications that need speed set as well as a servo-type controller.

10.3 The design of the system

The main components of the system are the accelerometer sensor, the nRF transceiver, and the DC motor. The accelerometer is basically a motion sensor that, in this study, functions to detect the user's head motion. And thus the receiver circuit must be attached to some sort of headpiece assortment. There are several types of accelerometer sensors that work by different principles, resistance and capacitance being the primary ones; the one implemented in this study is the MEMS capacitive accelerometer sensor, which, as the name implies, is based on capacitive properties, namely variable capacitance. The ADAXL335 accelerometer component detects the motion and gives an analog output which is proportional to the motion of the user's head. This signal will then be fed into the Arduino Uno microcontroller board through the input pins. At this point, the Arduino Uno converts the analog input signal into a digital signal and processes it by eliminating some noise to yield a higher resolution signal. After this, the signal will be assigned to the appropriate output pins to be fed into the nRF transmitter sensor. The nRF transmitter sensor will pick up this digital signal and wirelessly transmit it to the nRF receiver sensor that is connected to the receiver circuit, which will eventually control the electrical-mechanical

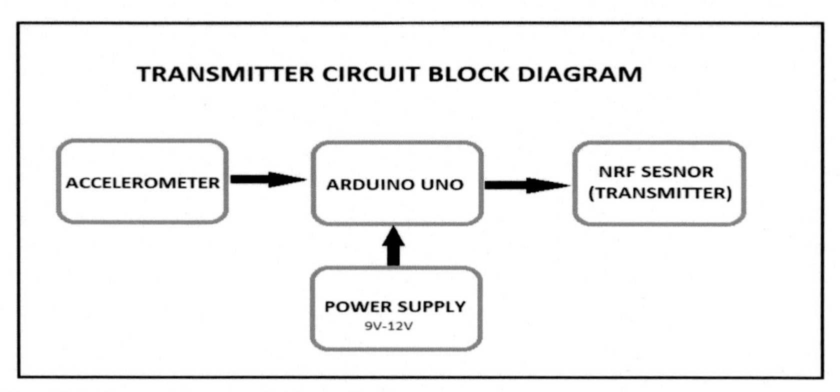

Fig. 10.4 Transmitter circuit block diagram.

wheelchair prototype. This circuit will require a 9–12 V power supply to power the Arduino Uno and is quite simple to set up, as demonstrated Fig. 10.4.

The received signal is processed by the Arduino and fed into the breadboard, which connects the relay and transistors to the controller board. The breadboard meshes all the components together: the relay, the transistors, and the RC controller board. These components allow us to operate the motion of the wheelchair using the signals transmitted from the first circuit as an input, thus eliminating the need for manual switches that would completely defeat the purpose of our study. The relay is composed of a coil holding a common armature between two other conducting plates (Nc and No) that complete the circuit. The principle behind it is based on electromagnetic properties. When a high voltage is fed through it, the armature is displaced as a result of the electrically induced magnetic field toward the No plate, which will connect the circuit and allow passage of current [25]. On the other hand, when insufficient voltage is fed through this component, the arm will not move and, in turn, will act as an open switch.

10.3.1 Limitations

The forms in which the motors in the controller board are set up limit the range of motion of the wheelchair. There are four motors, each of which drives one of the four wheels. The front two motors drive the forward motion. The backwards motion is driven by the rear pair of motors as shown in Fig. 10.5. For the rotational motion, a diode is placed between one side of the front and back motors, and only the front motor will operate to drive

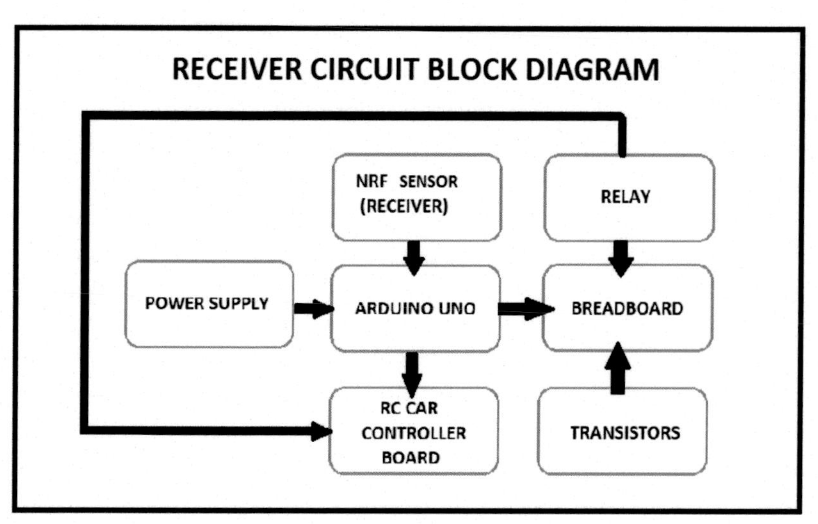

Fig. 10.5 Limitations and receiver circuit block diagram.

either a clockwise or counterclockwise motion. The disadvantage of this setup is that it is incapable of yielding a forward/backward motion and a rotational motion simultaneously.

When charging a battery, it is very important to pay attention and to not overcharge because overcharging may lead to side reactions (other than reactions that convert chemical energy into electrical), and these reactions occur in the electrical solution and lead to the conversion of some components of the solution to the gas that comes out of the battery. Overcharging the battery in terms of the end result will reduce battery performance and efficiency. When using a voltage regulator, we must check the output voltage on the multimeter before connecting the circuit so that a device (5 V) does not explode as a result of pushing a large voltage by mistake. More heat is generated by the regulator when a large volt is inserted, in which case the electrical circuit needs a heat sink to ensure that the electrical circuit is not burned.

10.3.2 Coding

The Arduino Uno compiler is based on C/C++ programming languages, which are basically the universal coding language that can be built on here. The coding required for this study was done by implementing previous knowledge from computer engineering courses at Near East University as well as with the help of some of the doctors and professors from the

Computer Engineering Department. To initiate this code, a plan was put in place to outline the basic working principle behind the code: what functions will the code carry out, how will it be represented, how will the physical components be linked to binary instructions to achieve the expected mechanism, what is the general layout of the code, and what is the order of instructions in the code? All of these are essential ideologies that need to be carefully planned and thought about in order to design the optimum code for our system. As previously mentioned, this study implements two independent Arduino Uno units for each of the electronic circuits: the transmitter and receiver circuits. For the transmitter circuit, the main concept behind it is that the MEMS accelerometer yields a readable or transmittable binary signal that can be used to drive the second circuit via the nRF transmitter device. The microprocessor in this circuit functions to read the MEMS accelerometer signal, convert it to a digital signal, perform noise reduction algorithms, and then feed this signal into the nRF component as an input. From the physical hardware perspective, the input/output pins positions are crucial for code assembly as we will define each one to represent the appropriate signal. For example, the MEMS accelerometer will connect to the Arduino Uno through one of the pins that will later be defined as input 1. For the receiver circuit, the coding gets a little more complex. The concept behind this circuit is to translate the signal received from the MEMS accelerometer into motion instructions to drive the motors of the wheelchair. For each signal a different range of motion is excited, so the motors activated will also differ; for example, forward motion excited by forward head movement, picked up as a signal from the MEMs accelerometer component, should activate the front two motor pairs that will accelerate the front two wheels of the wheelchair and cause an overall forward motion. These are coded in C programming language in the code of the Arduino Uno. Some main libraries of the Arduino code used in this study:

- SPI.h: helps in communicating with (serial peripheral interface) SPI devices nRF24L01.h; it is used for handling the modem driver;
- RF24.h: helps the users to control the modem radio;
- RF24 radio [26, 27]: we created an object called radio which represents the modem connected to the Arduino; signal CE is connected to the pin number 7 while signal CSN is connected to pin number 8;
- Rxaddr: a global array in this array; we define the modem address whose main function is receiving the data from the Arduino Uno; the address value is 00001;
- radio.begin(): responsible for activating the modem;

- radio.openReadingPipe(0, address): determines the modem address whose function is receiving data; the first argument represents the stream number while the second argument is the address;
- radio.startListening(): receiving data from the modem using this method;
- pinMode: defines the digital pins either as an output pin or as an input pin;
- Voidloop: allows the loop function, runs over and over again;
- digitalWrite: it is used to write high or low value to the digital pin after assigning it as an output pin;
- analogread(): its main function is to read the value of the along pin; the Arduino board has a multichannel 10-bit ADC that will convert the voltage, which is between 0 and 5 V, into integer values; this value ranges between 0 and 1023.

10.4 Conclusion

The research resulted in the design development of an advanced electric chair for the disabled at a reasonable price. These developments include adding mechanical parts to the chair to convert its movement from traditional manual movement to mechanical movement at a good degree of performance. Ninety percent of the users that tested the system rated it as 9.5/10. As previously mentioned, the aim of our study was to facilitate a better quality of life for those with disabilities, namely any disability that immobilizes muscular functions from the neck and down. We centered this study on head gesture by allowing the user to have access to a set range of movements. The present invention relates to a wheelchair that is controlled by the movement of the head. The system has an electric wheelchair frame, head movement signals originating from a head device, built-in movable head device with electrodes, MEMS accelerometer sensors, a signal unit and signal processing unit, a radio frequency transmission unit, a radio receiving unit, a microcontroller, a motor-powered unit, a motion-sensitive wheel control device that is ready to connect to the head of the user, a communication interface device to communicate with the wheelchair console, and a track tracker installed in an electric wheelchair to avoid obstacles. In summary, the proposed device has the ability to sense the user's head movement and translate it to wheelchair motion. The wheelchair can accelerate in three ranges of motion: forward motion, backward motion, and rotational movement. Future targeted research would be to expand the study with more advanced technologies, such as artificial intelligence and advanced detection mechanisms, to yield a state-of-the art product that can eventually be introduced into the market for public use. Our vision is to integrate advancing technologies with current in-use wheelchair models to yield a more efficient, functional, and user-oriented prototype that can eventually be of use for the target population. The main focus when developing this prototype was the user: what are the day to day hardships of patients who suffer from physical disabilities, namely paralysis and muscle detrition? How can we eliminate those hardships and limitations and allow for a more independent lifestyle? The intuition presented by this study is to subjectively improve the lifestyle of those who live with physical disabilities in a substantial manner.

References

[1] S.E. Cahill, R. Eggleston, Reconsidering the stigma of physical disability: Wheelchair use and public kindness, Sociol. Q. 36 (4) (1995) 681–689. s.l.
[2] E. Jeppsson Grassman, I. Holme, A. Taghizadeh Larrson, A. Whitaker, A long life with a particular signature life course and aging for people wih disabilities, J. Gerontol. Soc. Work 59 (2012) 95–111. s.l.
[3] L.R. Moll, C.A. Cott, S. Nixon, Qualitative evidence in chronic, disabling conditions (childhood- or early-onset physical impairment), in: Handbook of Qualitative Health Research for Evidence-Based Practice, Springer, New York, NY, 2016.
[4] D. Labbe, W. Ben Mortenson, P.W. Rushton, L. Demers, W.C. Miller, Mobility and Participation Among Ageing Powered Wheelchair Users: Using a Lifecourse Approach, Ageing & Society, Vancouver, Canada, 2018, pp. 1–17.
[5] R.C. Simpson, Smart wheelchairs: A literature review, J. Rehabil. Res. Dev. 42 (4) (2005) 423–436. Pittsburgh, PA.
[6] R.A. Cooper, Engineering manual and electric powered wheelchairs, Crit. Rev. Biomed. Eng. 27 (1–2) (1999). s.l.
[7] A.J. Rentschler, R.A. Cooper, S.G. Fitzgerald, M.L. Boninger, S. Guo, W.A. Ammer, D. Algood, Evaluation of selected electric-powered wheelchairs using the ANSI/RESNA standards, Arch. Phys. Med. Rehabil. 85 (4) (2004) 611–619. s.l.
[8] L.H. Van der Woude, S. de Groot, T.W. Jassen, Manual wheelchairs: reasearch and innovation in rehabilitation, sports, daily life and health, Med. Eng. Phys. 28 (9) (2006) 905–915. s.l.
[9] D. Ding, R.A. Cooper, Electric powered wheelchairs, IEEE Control Syst. Mag. 25 (2) (2005) 22–34. s.l.
[10] R.D. Ball, Pediatric wheelchair, 5,592,997, US, 1997.
[11] W.G. Gurasich, M.H. Bussel. Othotic wheelchair positioning device and support system, 5,551,756 US, September 3, 1996.
[12] S.N. Eichenholtz, Management of long-bone fractures in paraplegic patients, J. Bone Joint Surg. Am. 45A (1963) 299–310. s.l.
[13] A.S. Abramson, Bone disturbances ininjuries to the spinal cord and cauda equina (paraplegia), J. Bone Joint Surg. Am. 30A (1948) 982–987. s.l.
[14] R.K. Sheilds, Muscular, skeletal and neural adaptions following spinal cord injury, J. Orthop. Sports Phys. Ther. 32 (2002) 65–74. s.l.
[15] Y.A. Badamasi, The working principle of an Arduino, in: 11th International Conference on Electronics, Computer, and Computation (ICECCO), 2014, pp. 1–4. s.l.

[16] B.M. Al-thobaiti, I.I. Abosolaian, M.H. Alzahrani, S.H. Almalki, M.S. Soliman, Design and implementation of a reliable wireless real-time home automation system based on Arduino Uno single-board microcontroller, Int. J. Control Autom. Syst. 3 (3) (2014) 11–15. s.l.
[17] L. Louis, Working principle of Arduino and using it as a tool for study and reasearch, Int. J. Control Autom. Commun. Syst. I (2) (2016), https://doi.org/10.5121/ijcacs.2016.1203. Ahmedabad, India.
[18] D. Mellis, M. Banzi, D. Cuartielles, T. Igoe, Arduino: An open electronic prototyping platform, in: Prc. CHI, vol. 2007, 2007. s.l.
[19] L.R. Hinze, Fastenerless sealed electronic module, 5,70,754, Washington DC, US, December 30, 1997.
[20] N. Yazdi, F. Ayazi, K. Najafi, Micromachined intertial sensors, Proc. IEEE 86 (1998) 1640. s.l.
[21] S. Narayanan, Synchronization of Wireless Accelerometer Sensors for Industrial Application, 2019.
[22] B.A. Khan, et al., Implementation of micro controller based electromechanical over current relay for radial feeder protection, in: International Conference on Engineering and Emerging Technologies (ICEET), 2019, pp. 1–6. s.l.
[23] J. Kunal, A. Siddhartha Rao, Over current protection of transmission line using GSM and Arduino, Int. J. Eng. Trends Technol. 50 (1) (2017). s.l.
[24] L.S. Robertson, Safety belt use in automobiles with starter-interlock and buzzer-light reminder systems, Am. J. Public Health 65 (12) (1975) 1319–1325. s.l.
[25] O. Koçak, E. Gurel, A. Alpek, A. Kocoglu, Control of wheel chair for quadriplegia patients: design a bioremotecontrol, in: 9th International Conference on Electronics Engineering, 2015. s.l.
[26] B. Woods, N. Watson, A short history of powered wheelchairs, Assist. Technol. 15 (2) (2003) 164–180. s.l.
[27] Chair Institute. (Online) January 02, 2020. https://chairinstitute.com/types-of-wheelchairs/ (Cited 2 June 2020).

CHAPTER ELEVEN

Construction of vehicle shutdown system to monitor driver's heartbeats

Dilber Uzun Ozsahin[a,b,c], John Bush Idoko[d], Basil Bartholomew Duwa[b], Majd Zeidan[b], and Ilker Ozsahin[a,b,e]

[a]DESAM Research Institute, Near East University, Nicosia, Turkish Republic of Northern Cyprus, Turkey
[b]Department of Biomedical Engineering, Near East University, Nicosia, Turkish Republic of Northern Cyprus, Turkey
[c]Medical Diagnostic Imaging Department, College of Health Sciences, University of Sharjah, Sharjah, United Arab Emirates
[d]Applied Artificial Intelligence Research Centre, Department of Computer Engineering, Near East University, Nicosia, Turkish Republic of Northern Cyprus, Turkey
[e]Brain Health Imaging Institute, Department of Radiology, Weill Cornell Medicine, New York, NY, United States

11.1 Introduction

When the heart is faulty, it leads to ineffective living, and if not detected early, it can lead to death. When the heart muscles become weakened, the heart enlarges and pumps less effectively [1]. Due to this ineffectiveness, it could lead to swollen legs and feet, breathlessness, cough, paroxysmal nocturnal dyspnea, and congestion of blood. It is advised to take good care of the heart and to avoid anything that can could lead to heart failure or diseases. Some of the causes of heart failure are high blood pressure, leaking of narrowed heart valves, slow or fast heart rhythms, anemia, thyroid disease, and lung disease [2]. Heart failure could also be caused by a human factor, and this includes alcohol, drugs, and viral infections. Here are diseases related to the condition of the heart:

- Congenital heart disease: It is a disease that has to do with the structure of the heart. It is usually the most common type of birth defect.
- Arrhythmia: This could affect the rate or rhythm of a person's heart. The heart could beat too quickly, too slowly, or with an irregular pattern.
- Coronary artery disease: This has to do with the narrowing or blockage of the coronary arteries, which is usually caused by atherosclerosis. Atherosclerosis is the hardening or clogging of the arteries. It is a build-up of cholesterol and fatty deposits called plaques on the inner walls of the arteries.

Modern Practical Healthcare Issues in Biomedical Instrumentation
https://doi.org/10.1016/B978-0-323-85413-9.00013-X

- Dilated cardiomyopathy: This is the disease of the heart muscle. The ventricles stretch and dilate and cannot pump blood as well as a healthy heart can.
- Myocardial infection: It is known as heart attack. This happens when blood flow reduces or stops to a part of the heart, causing damage to the heart muscle. Symptoms include chest pain or discomfort in which the pain may move into the shoulder, arm, back, neck or jaw.
- Heart failure: This is when the heart is unable to pump sufficiently to maintain blood flow. Symptoms could include shortness of breath, excessive tiredness, and leg swelling.
- Hypertrophic cardiomyopathy: This occurs when part of the heart becomes thickened without any cause, and this makes the heart unable to pump blood effectively. Signs include feeling tired, leg swelling, and shortness of breath.
- Mitral regurgitation: This is an abnormal reversal of blood from the left ventricle to the left atrium. It is caused by disruption in any part of the mitral valve apparatus [3–5].

The heart is an important organ in the human body. Hence it is important to take care of our heart, and this can be done by carrying out physical exercises, eating healthy, drinking a lot of water, consuming fruits and vegetables, eating fish rich in omega-3-fatty acid, reducing salt intake, and avoiding smoking. Regular check-ups should be done yearly to determine the state of the heart. However, the heart could be deformed from birth, or the heart valve could be diseased, hence causing a faulty heart or leakage of blood in the heart [6].

The increase in heart failure is threatening to the global health services and financial institutions due to the global increase in the mortality rate. In Egypt, Attorney General Nabil Sadeq revealed that the cause of the fatal train accident that took place at the main train station in downtown Cairo and that killed and injured dozens was a result of a heart-related illness [7]. Heart failure is a complex clinical syndrome and has a high rate of mortality and morbidity in the United States and globally. It is expected that by the next decade, the diagnosis of individuals with heart failure will increase by 46%. Furthermore, patients with heart failure often have additional cardiovascular and noncardiovascular comorbidities, such as arrhythmias, obstructive sleep apnea, diabetes, and depression, that complicate care [8].

Due to the causes of heart failure in individuals, techniques and approaches have been developed to help reduce the death rate of people with heart disease. From research, it has been shown that people with heart failure may not drive due to fear of death or accident. In this project, a device is developed to monitor and control the heartbeat of an individual who has heart-related diseases to avoid accidents while driving with the use of sensors and alarm systems. This system automatically alerts the individual when his pulse rate is high, and in some emergency cases, the alarm systems completely shut down the vehicle [9].

11.2 Literature review

In this literature review, there is discussion of how dynamic events (conscious or unconscious error) can affect a person's heartbeat due to (false) perception of ability and lack of attention. An experimental study was done where a watch was used to monitor the heartbeat of an individual, and it concluded that there were errors (to study the perception abilities when the occurrence of alarm is synchronized or desynchronized with heartbeats) and that a better analysis of the alarm should be used to check a patient's heartbeat [10]. Another review describes how automated systems are used to monitor a driver's heartbeat; the study is aimed at building a driver mental state model that can determine the quality of the drivers' responses in a highly automated vehicle. The study was effective, readings by a driving simulator measuring the driver's response time in seconds for a takeover request. A high response time by a driver correspondingly leads to a poor takeover [11]. The interatrial shunt device is a nitinol device with a 19 mm outer diameter and a central opening diameter of 8 mm designed to sit across the interatrial septum; it is used for patients with heart failure, but a more durable solution and therapeutic strategies are needed for effective results [12]. Durable circulatory support with a paracorporeal device as an option for pediatric and adult heart failure; this device is also used for patients with other heart diseases, but the disadvantage is that it could be tampered with by thromboembolism, mechanical problems, and infections [13]. Angiotensin receptor neprilysin inhibitors (ARNI) can improve blood pressure and decrease diuretic requirement and functional capacity in left ventricular assist device (LVAD) patients. LVAD speed and medication adjustments may be necessary following ARNI initiation, but side effects can limit the ability to sustain ARNI therapy in the setting of LVAD [14].

11.2.1 Statistics of the sudden death

In the United States, approximately 1.4 million people ages 1–34 dies every year, and two percent of those can be categorized as a sudden cardiac death, an unexpected event for individuals within that age range [15].

11.2.2 Sudden death types and causes of each type

11.2.2.1 Sudden cardiac death

As previously mentioned, the SCD is a loss of the heart function of receiving the blood from the different body parts, sending it to the lungs to fill it with oxygen, then receiving it again from the lungs as oxygenated blood before resending to the body parts; this is the most important mission that happens inside of humans and animals, and if it is stopped for a few seconds, the body will not get its oxygen, which it needs in every second to make the different biological activities happen, so from this point, we know that if the heart function stops for a few seconds then that will lead to a sudden and immediately death for the whole body. A heart attack differs from an SCD as a heart attack is caused by a blockage that occurs inside of the vessel track that can prevent the movement of blood, but it is an important cause that may lead to an SCD, written about with detail in Section 11.2.2.2.

11.2.2.2 Causes of the SCD

The reason that we are unable to find a definite cure for this SCD based disease is because it has many different causes and possibilities, and it doesn't have an expected time or age to occur in order to suggest preventative measures for avoiding SCD altogether. The most common reason for SCD is any problem that may happen in the cardiac normal rhythm, such as arrhythmia, which is supposed to happen regularly by an electrical pulse that starts at the node of the upper side of the heart and spreads along specified parts to the lower base of the heart muscle (apex of the heart). There are many other causes, as we mentioned, that are not direct causes like arrhythmia, but there are other diseases that can occur and lead to SCD. Such diseases include coronary artery disease, which happens because of an excessive amount of cholesterol in blood; a heart attack (mentioned in Section 11.2.2.1); an enlarged heart, which can cause an irregular rhythm; valvular heart disease, such as leakage of blood from a certain heart valve; congenital heart disease, which is caused by a heart defect that a person has likely had from birth; and electrical problems in the heart, like the electric pulse, not spreading as needed.

11.2.2.3 Internal bleeding of the head

Internal bleeding is a very dangerous case in all of its types, but internal bleeding of the head will be discussed because it is known as one of the prominent causes of a sudden unexpected death. Internal bleeding in other parts of the body may cause a bug harm inside of the body's tissues, but they do not usually cause sudden death to an injured person.

An intracerebral hemorrhage (ICH) is a type of stroke caused by bleeding of small blood vessels in the brain, which occurs as a result of high blood pressure in the small arteries that supply the brain with blood, nutrients, and oxygen, so hypertension causes the rupture of the thin walls of these arteries, leading to hemorrhages within the brain; and as a result of the accumulation of blood clotting and fluid, the pressure on the brain increases, which may lead to the compression of the brain or an aneurysm. As artery walls rupture, the area fed by the artery becomes deprived of oxygen. The brain may be pushed down by pressure from the small hole in the base of the skull called the foramen magnum and the parts of the brain that come into contact with the bone" should be removed, it adds no value. Thank you for the observation. The symptoms of internal bleeding in the head may vary depending on the location, severity, and size of the bleeding. Symptoms may develop suddenly or over time, and they may worsen gradually or suddenly. Brain hemorrhage is an emergency medical condition that requires immediate medical care.

Symptoms of internal bleeding in the head include suffering from acute and sudden headaches; tinnitus seizures, for the first time without prior history of seizures; weakness in arm or leg; nausea or vomiting; hibernation and lack of vigilance; changes in vision; tingling or numbness; difficult speech, or difficult to understand speech; difficulty swallowing; difficult to write or read; loss of fine motor skills, such as hand tremors; loss of balance; unusual sense of taste; and unconsciousness. Internal bleeding in the head can be caused by one of the following reasons: head trauma, such as falls, car accident, sports accidents, etc. Hypertension causes rupture of vessel walls. An artery in the brain is blocked by a blood clot formed in the brain or transmitted to the brain from another part of the body, leading to leakage of blood from the damaged artery, rupture of dilated cerebrospinal vessels, the presence of a weak point in the wall of the blood vessels causing its explosion, and the accumulation of amyloid protein within the artery walls of the brain, known as cerebral amyloid angiopathy. Hemorrhage of the arteries or veins that are formed abnormally, known as arteriovenous malformation, are largely affected by anticoagulants such as blood thinners; smoking or excessive drinking; drug abuse, such as cocaine; and pregnancy or childbirth complications, including eclampsia or postpartum vasculopathy. An intracerebral hemorrhage/intracerebral aneurysm in the head can be categorized into several types: epidural hematoma, subdural hematoma, subarachnoid hemorrhage, and intracerebral hemorrhage.

Epidural hematoma, a blood tumor is defined as a group of blood clotting outside a blood vessel. This is caused by the accumulation of blood between the skull and the outer capsules of the brain. Most cases of epidural hemorrhage result from a head injury with a skull fracture. The epidural tumor is characterized by short-term loss of consciousness and subsequent recovery. Subdural hematoma is known as the accumulation of blood on the surface of the brain and usually occurs as a result of the movement of the head quickly forward and stop. This type of bleeding in the head is more common in the elderly and in people who consume large amounts of alcohol. Subarachnoid hemorrhage occurs as a result of hemorrhage between the brain and thin tissue covering the brain, known as meninges. Trauma is the most common cause of this type of hemorrhage in the head and may occur because of the rupture of one of the major blood vessels in the brain, such as intracerebral aneurysm. This type of hemorrhage is characterized by sudden acute headache that precedes this type of bleeding. Symptoms also include loss of consciousness and vomiting. With an intracerebral hemorrhage, bleeding occurs within the brain due to cerebral artery aneurysms burst and is the most common bleeding in the head. A prominent warning sign for this type of bleeding is the sudden onset of neuropathy.

11.3 Driver alertness detection system

Research and development centers of car companies have begun to devise new safety techniques in recent years aimed at preventing drivers from drowsiness or sleeping while driving, especially since many international studies have shown that about 20% of road accidents are caused by drivers' drowsiness/sleeping while driving. The reason can be fatigue and exhaustion or tiredness and lack of sleep; studies have reached that that possibly up to 50% of fatal accidents involve these same reasons. The most advanced security technology is the Drowsiness Detection System. The Drowsiness Detection System is designed to prevent accidents by protecting drivers from feeling drowsy while driving because such systems have the ability to identify and analyze drivers' driving patterns constantly. This system is present in the category of positive safety measures that prevent accidents before they occur, compared to negative safety systems that reduce the risks and consequences of accidents as they actually occur.

These technologies have evolved over time to include a large list of advanced systems that are rich in modern cars, and the most prominent of these techniques are available at the present time:

Routing mode control system: This system is mainly used to monitor the driver's steering input of the steering wheel via the power steering system and to identify any sudden or unwarranted change in the usual steering pattern.

Traffic lanes control system: This system is based on an external camera that monitors the traffic lanes and senses any sudden change within those lanes, including any drivers unjustifiably exiting the traffic lane, to alert the driver both visually and by voice to correct the route.

Eye/face control system: This system requires the presence of a special camera inside the car in front of the driver, monitoring the face or eyes of the driver to determine the variables associated with drowsiness, such as the fall of the head down, closing of the driver's eyes, and other variables.

Systems available in the automotive industry: Driver Alertness Detection System (DADS) is a system available to many large companies, and it has the ability to detect the level of alertness of drivers to notify them of them of any change.

- The Audi system advises the driver to get a Rest Restriction System.
- Active Driving Assistant: This system is accompanied by a sensor system to support the Attention Assistant in jointly analyzing the driver's behavior on a continuous basis and advising the driver when necessary to stop to get rest through the central display inside the car in the driver's face.
- Bush's drowsiness control system: This receives inputs from special sensors that monitor the angle of the steering wheel along with a front camera to monitor the traffic lanes and other inputs including the speed of the car.
- Driver Alert by Ford: Ford unveiled this system in 2009 and was the latest innovative technology in this area. It depends on monitoring drivers' stress levels through many of the inputs and issues a warning voice and visual alert to drivers. The system is also linked to the navigation and guidance technology available in the car to notify the driver of the nearest places on the road for coffee/relaxation.

Antisleep pilot: This is a Danish-made device that can be installed inside any car, and it relies on motion sensors, steering, and reflexes to alert the driver.

Vigo: It is a Bluetooth-connected headphone that recognizes any signs of sleepiness through head and eye movement and uses a combination of lighting, sound, and vibration to alert drivers when necessary.

Systems Applications and Products (SAP) Enterprise in Data Processing: SAP developed a facial recognition system integrated into car system to determine the exposure of drivers to drowsiness. The system uses special night vision cameras that analyze the movement of the facial muscles and the rate of closing the eye.

It is worth mentioning that there are many other systems available in modern cars that indirectly target driver alerting while driving, including LED lights with dazzling lights designed to keep the drivers awake while driving at night. There are also other systems available on the same roads, including the traffic lane system, which relies on a special material placed on the lane lines to produce vibration and high sound of the tires in case the driver exits the traffic lane [14].

11.3.1 Blind spot detection

This technique uses sensors or a camera under each of the two side mirrors that are constantly working. To scan this area that you cannot see from the driver's seat, it is very clear with the use of a camera or sensor but can be difficult while driving; if there is a large object, it lights the LED lights on the rear mirror to warn you of a vehicle in this area. In some cases, however, these sensors may be subject to deception. They may fail to give timely warning according to their quality. Imagine a speeding car coming from behind; at the last minute, the driver decides to change course. In this case, the system will certainly not warn at that time. It is appropriate for a driver to decide to change course in this context.

In addition, some systems cannot monitor nearby motorcycles or bicycles, two of the most dangerous vehicle types. Of course, we do not say that these systems are not useful, but you should pay attention to both sides of your car even if the blind area light is extinguished. There is also the active blind area control system, which moves the vehicle back on track if a vehicle is detected in the hidden area. While it may seem encouraging, the same system is integrated with one of the previous "maintain path" systems, so it may have the same disadvantages.

11.4 Methodology

In this section, we are going to explain in detail the components that are used in this project as well as discussing the exact need of each component to make the project work well. Then, we will discuss how we developed it using simple steps with a circuit image of real connections, and lastly, there will be discussion of the working principle of the device.

11.4.1 Materials and methods

(A) Arduino Uno circuit

It is a microcontroller board chip that is programmed by software for receiving and sending signals.

The Arduino Nano board is a microcontroller board that's designed with 14 digital pins, 8 analog pins, and a USB connection based on the ATmega328P microcontroller chip (Table 11.1).

In this paper, it serves as the main controller, or brain, of the device as it receives signals, processes them, and sends them, if needed.

The MAX30100 is an incorporated heartbeat oximetry and heart rate measurement; it can also calculate the O_2 percent in the oxygenated hemoglobin (HB). It is a board made using the sensor MAX30100 to simulate the real hospital using a pulse oximeter. It is working using power supply from 3 to 5 V. To use the board the best way, we did so by removing the board welded resistors and adding outer resistors, almost with the same resistance ohms, because the designed board wouldn't work well by the Arduino if we didn't make this move.

Table 11.1 Specifications of the Arduino Nano board.

Chip	ATmega328P
Supplying	5 V
I/P supply	7–12 V
Digital slots	22 (of which 6 provide PWM output)
Pulse width modulation digital slots	6
Alternating input slots	8
The direct current	40 Ma
SD	32 KB (ATmega328P) Of which 2 KB used by bootloader
Static random access memory	2 KB (ATmega328P)
Electrically erasable programmable read-only memory	1 KB (ATmega328P)
Clock speed	16 MHz
Weight	7 g

As written by us, this sensor job in the circuit was the main sensor to detect the existence of heartbeat and give the result to the Arduino Nano board to make the next move

(B) Pulse oximeter

A servomotor is a small, simple, and useful motor that can be used with many orders because of its high and great precision of reaching the exact same angle, such as 90 degrees, 180 degrees, etc. It is working according to its type; so if it is a DC motor, then it needs a DC power supply. It is calling a DC servomotor, the same if it was AC type; then, it would need an AC power supply and could be called an AC servomotor. The best advantage of this motor type is that it is small and lightweight. The position the motor takes depends on the coding order, which will be coming in this project from the Arduino Nano board. The servomotor used in this project is the main car motor, and we used two motors in a way to show the wheels for both sides of the car.

(C) Breadboard

This is where the main connection is done. It consists of holes, and electronic components can be inserted into it. It is an electrical board that plays the role as a connection board; it has many shapes and sizes according to its use. Each five points in a row are connected together from the internal side to allow us to plug in a pin of any component anywhere of these five points and connect the other pin component, which must be connected to the first pin in another point of them, so they will be connected easily. It also has two long columns at its terminals that are all connected together on that column, from the inside, to allow the main board supply from each two points as negative and positive sides for supplying the whole circuit components. We used our breadboard as the main board to connect our circuit by it, and it was much simpler because it is a solderless board; so whenever we want to change something in the circuit, it will be much easier to do.

(D) Resistors and transistors

The wires are made of cobber, and they are designed small to fit the size of Arduino projects; so it's just used to connect one electric signal from a place to another in the circuit. The resistors are designed to protect the circuits from residual voltages that may pass across it, and it's used for the same reason in this project. They are used as different ohms in accordance with the electrical or electronic component that they are protecting. Here we used the wires to move the signals between the components of the circuit from any place to another by connecting their terminals in the appropriate way. The resistors are used to protect and improve the MAX30100 board efficiency. The full circuit diagram of the novel technology is depicted in Fig. 11.1.

The vehicle shutdown device uses a power supply that generates electricity to the Arduino circuit where it programs and sends signals to the pulse oximeter. The pulse oximeter picks the signals and detects if there is change in the heartbeat before alerting the two servomotors for the ON/OFF signal.

11.5 Device software simulation

As a simulation for our device and to make sure that it will be working correctly before connecting it, we used the Proteus software program that contains most of the main electrical and electronic parts that are available on the market, and the Arduino was available for use and to be added to the circuits at any time. The programming of the propose system is performed using C language. We connected the device and tested it in the Proteus, and afterwards, we ensured that it was working well in the simulation; we designed it in the real-world hardware, and the programming was made by the Arduino software program, which can be easily downloaded. After adding some libraries, we were able to upload our programing code to the Arduino Nano.

11.5.1 Working principle

Firstly, the device will be plugged into the power bank that is attached as a power source, and the part that will directly receive the power from the power bank is the Arduino Nano, which in turn will provide the power to rest of the circuit components because, as mentioned before in

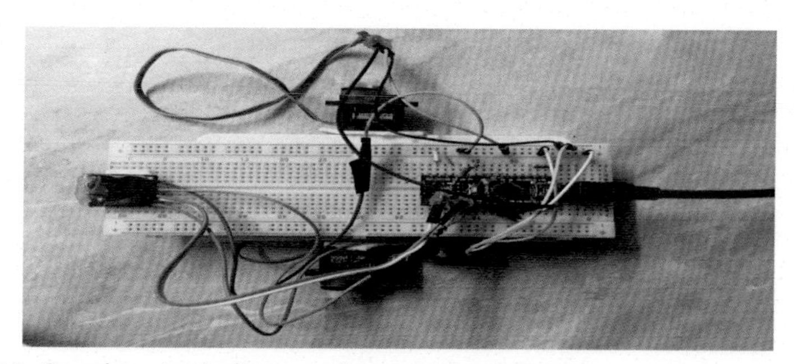

Fig. 11.1 Complete circuit diagram of the proposed system.

Section 11.4.1, there are two pins of the Arduino Nano designed to provide 5 and 3.3 V to the parts requiring them; also, there are two pins designed as grounds to close the circuit. Then, after the whole circuit components powered and turned on and after a little while, the servomotors will reset themselves to zero degree in preparation. After that, the sensor will take almost ten seconds to prepare for use, and as long as there was no finger attached to the surface of the pulse oximeter sensor, the motors will stay at their zero level. Then, a person will be asked to touch the sensor with his finger to test the response; as a result of finger placement on the sensor, in about 5–10 s the motors will be working as they are programmed. They will keep going fast to 90 degrees, 180 degrees, and back to zero in a continuous manner, and both of them will be moving to the same direction simultaneously to guarantee that the result is accurate enough; there will be no problems with motor response time. Lastly, as a test for the whole system, the finger will be removed, and after less than 5 s, the motors will stop together at the zero level. If we further need to test the system, we can place the finger on the sensor again and remove it as many times as needed to measure the response time of each case.

11.6 Results

The vehicle shutdown device is significant because it helps to prevent accidents that could cause death and serious injuries. This device is helpful to individuals with heart diseases or those with high blood pressure because it helps them to monitor and control their pulse heartbeat. Electronic components are incorporated into this one device that is attached to the vehicle. When the device is powered, it begins to carry out its function, and if it notices any change in the pulse rate of the individual, the vehicle will shut down.

11.7 Discussion

As we see in Table 11.2, the device's motors are not working when there is no heartbeat detected by the pulse oximeter MAX30100, and if there was a heartbeat present, the normal heart rate is from 60 to 100 bpm. The motors are just beginning to work when the heartbeats are detected and in the normal range because, as mentioned in the second section, the abnormal heart rate may be a sign of a heart disease. The device is working well now and as programmed. It is low-cost with simple electronic components that are available on the market, serves as a highly effective and

Table 11.2 Results of the test.

Heartbeats (bpm)	Servo1	Servo2	Angle (degrees)
0	OFF	OFF	0
10	OFF	OFF	0
20	OFF	OFF	0
30	OFF	OFF	0
40	OFF	OFF	0
50	OFF	OFF	0
55	OFF	OFF	0
59	OFF	OFF	0
60	ON	ON	0–90–180
70	ON	ON	0–90–180
80	ON	ON	0–90–180
90	ON	ON	0–90–180
99	ON	ON	0–90–180
100	OFF	OFF	0
105	OFF	OFF	0
110	OFF	OFF	0
120	OFF	OFF	0
130	OFF	OFF	0

accurate device that features continuous detection, saves lives by preventing cars from moving without a driver behind the wheel, decreases the rate of car accidents by prevention, has a proper small size that allows for optimal placement, and delays action to prevent unwanted consequences. The device will be inserted in a car that is generally moving, so the vibrations that happens while the car is moving may cause interference in the device readings; if the driver was playing some sports before he gets into his car, this will increase his heart rate temporarily, so the car's motor will not turn on until he has reached a rest condition, possibly causing an inconvenience for the driver (Table 11.3).

Table 11.3 Normal heart rate ranges for different ages.
Acceptable ranges of heart rate

Age	Heart rate (beats per minute)
Infant (6 months)	120–160
Toddler (2 years)	90–140
Preschooler	80–110
School-age	75–100
Adolescent	60–90
Adult	60–100

11.8 Conclusion

There are different reasons why a person could experience an accident while driving, and one of them could be heart failure. Heart failure is a condition whereby the muscles are weakened; hence it requires pumping the blood at a high voltage, leading to high blood pressure. The heart failure could be chronic and caused by impaired left ventricular (contraction) or systolic heart failure. The symptoms of heart failure include breathlessness, cough, shortness of breath, and congestion of lungs and swollen legs. Heart failure can be avoided if we are careful of what we eat and drink. It is advised we eat clean by eating food that is less in calories, eating more vegetables, drinking plenty of water, and engaging in regular exercise. Some causes of heart failure include alcohol, because heavy consumption affects the heart, and anger, because when it is not controlled properly, it could also lead to heart failure. When a person decides to drive with influence from either alcohol or anger, it could also lead to an accident. People also experience road rage for so many reasons: lateness to work, tailgater, desire for speed, and petty driving behaviors.

All the major causes of accidents listed above can be controlled and monitored by a device that checks the pulse rate of individuals while driving. In this project, the device called is referred to as a vehicle shutdown device; it depends on the driver's heartbeat in controlling and monitoring the pulse. This helps to prevent accidents and ensure that those with heart failure could also drive safely without any fear or accident because about 46% of people who have heart failure or disease do not drive. This device serves as a monitor or detector to tell the individual when he or she needs to stop the vehicle, and sometimes, the device could also shut down the vehicle in emergency cases. With consideration of statistics on car accidents, we developed the idea of making a device that prevents or decreases the harm of accidents that may be caused by sudden death or heart disease.

According to our results, the device is working well as a prototype, and it is safe to be manufactured in real life. We picked up people with different genders and ages to compare if it was working well with all of them, and it did work; so when we tried the device on a person in the middle age and middle thickness of finger skin, the device was working as the motors are moving when the heart rate is normally more than 60 bpm and less than 100 bpm, which is the safe area with minimal risk, but when the device detects more than 100 bpm or less than a 60 bpm, the motors will stop working after a few seconds to ensure that the problem is still continuous and that there is no reading defects or problems. The device worked as designed, and it will improve with the application of future research and recommendations.

11.8.1 Future aspects and recommendations

This device can be more useful when connected to a wireless connector, so there's no need for the wires when it can be made with a Bluetooth module; the other medica-made types of pulse oximeters may be more accurate if they were used in such a device discussed here and possibly to prevent and decrease the amount of interference from the car's movements, which this sensor type is highly affected by during use. This device can also be extended to include many other types of diseases such the diseases that are caused by the low oxygenated hemoglobin amount because we can measure both heart rate and oxygen saturation with the same pulse oximeter in the hemoglobin, and it can be easily inserted somewhere in the car where the driver is usually putting his hand. Lastly, we recommend that when this device is used in real life, there must be more than one sensor inserted in the car to be able to know the heart rate wherever the driver puts his hand on the steering wheel.

References

[1] C.C. Beladan, S.B. Botezatu, Anemia and management of heart failure patients, Heart Fail. Clin. 17 (2) (April 2021) 195–206, https://doi.org/10.1016/j.hfc.2020.12.002.
[2] V. Muthurangu, Cardiovascular magnetic resonance in congenital heart disease: focus on heart failure, Heart Fail. Clin. 17 (1) (January 2021) 157–165.
[3] C. Laurence, M. Burch, Understanding heart failure: pathophysiology and approach to therapy, Paediatr. Child Health 31 (2) (February 2021) 61–67.
[4] K. Lau, A. Malik, F. Foroutan, T.A. Buchan, J.F. Daza, N. Sekercioglu, A. Orchanian-Cheff, Resting heart rate as an important predictor of mortality and morbidity in ambulatory patients with heart failure: a systematic review and meta-analysis, J. Card. Fail. 27 (3) (March 2021) 349–363, https://doi.org/10.1016/j.cardfail.2020.11.003.
[5] Y. Isler, A. Narin, M. Ozer, M. Perc, Multi-stage classification of congestive heart failure based on short-term heart rate variability, Chaos Soliton Fract. 118 (January 2019) 145–151.
[6] N. Silverstein, P. Wengrofsky, E. Botti, F. Oleszak, I. Bukharovich, Congestive heart failure readmission and guideline based therapy rates improvement at safety net hospital heart health center, J. Card. Fail. 26 (10, Suppl) (October 2020) S120.
[7] J. Olligs, D. Linz, D.G. Dechering, L. Eckardt, P. Müller, Heart rate—a complex prognostic marker in acute heart failure, Int. J. Cardiol. Heart Vasc. 26 (February 2020) 100456.
[8] A. Jovic, K. Brkic, G. Krstacic, Detection of congestive heart failure from short-term heart rate variability segments using hybrid feature selection approach, Biomed. Signal Process. Control 53 (August 2019) 101583.

[9] D. Pan, P. Pellicori, C. Walklett, A. Green, A.R. Masse, J. Wood, J. Purdy, A.L. Clark, Driving habits and reaction times on a driving simulation in older drivers with chronic heart failure, J. Card. Fail. 26 (7) (July 2020).

[10] F. Vanderhaegen, M. Wolff, R. Mollard, *Non-conscious errors in the control of dynamic events synchronized with heartbeats*: A new challenge for human reliability study, Saf. Sci. 129 (September 2020) 104814.

[11] M.T. Alrefaie, S. Summerskill, T.W. Jackon, In a heartbeat: using driver's physiological changes to determine the quality of a takeover in highly automated vehicles, Accid. Anal. Prev. 131 (2019) 180–190.

[12] S. Emani, D. Burkhoff, S.M. Lilly, Interatrial shunt devices for the treatment of heart failure, Trends Cardiovasc. Med. 3 (October 2020), https://doi.org/10.1016/j.tcm.2020.09.004. S1050-1738(20)30123-7.

[13] S.E. Bartfay, G. Dellgren, S. Hallhagen, H. Wåhlander, P. Dahlberg, B. Redfors, J. Ekelund, K. Karason, Durable circulatory support with a paracorporeal device as an option for pediatric and adult heart failure patients, J. Thorac. Cardiovasc. Surg. 161 (4) (April 2021) 1453–1464.e4, https://doi.org/10.1016/j.jtcvs.2020.04.163.

[14] K. Freed, R.L. Goldberg, B. Marino, R. Fioretti, N. Klemans, C.W. Choi, A. Kilic, R. Florido, K. Sharma, N. Gilotra, S. Hsu, Sacubitril-valsartan improves blood pressure and heart failure in left ventricular assist device (LVAD) patients, J. Card. Fail. 26 (10, Suppl) (October 2020) S40.

[15] G.L. Sumner, et al., Sudden cardiac death, in: Encyclopedia of Cardiovascular Research and Medicine, Elsevier, 2018.

CHAPTER TWELVE

Development of a modern electronic stethoscope

Dilber Uzun Ozsahin[a,b,c], John Bush Idoko[d], Busayo Oluwatobiloba Aderotoye[b], Laith M. Alasais[b], Hamdi Burakah[b], Jamil Hilal Seif Abu Shaban[b], and Ilker Ozsahin[a,b,e]

[a]DESAM Research Institute, Near East University, Nicosia, Turkish Republic of Northern Cyprus, Turkey
[b]Department of Biomedical Engineering, Near East University, Nicosia, Turkish Republic of Northern Cyprus, Turkey
[c]Medical Diagnostic Imaging Department, College of Health Sciences, University of Sharjah, Sharjah, United Arab Emirates
[d]Applied Artificial Intelligence Research Centre, Department of Computer Engineering, Near East University, Nicosia, Turkish Republic of Northern Cyprus, Turkey
[e]Brain Health Imaging Institute, Department of Radiology, Weill Cornell Medicine, New York, NY, United States

12.1 Introduction

Considering the simple user-friendly approach of the iconic stethoscope, the possibility of its obsolescence may, however, seem quite unlikely since there are many situations where the traditional stethoscope is capable of performing its jobs—including patient validity, assessment of pregnancy, identification of murmurs, determination of blood pressure, and diagnosing heart-related problems—in a world where adoption of newer methods or techniques are becoming popular due to the advancement in technology and products designation. The main objective of this chapter is to provide the details of the design and the systematic approach towards developing an electronic stethoscope capable of capturing, amplifying, and processing sounds that are produced in identified internal organs for analyses and visualization through viable digital signal processing methodology. This device optimizes hearing of filtered low-frequency sounds produced in the heart or lungs via a loudspeaker that allows for accurate medical diagnosis that could have been altered due to the low amplitude of the protruding signals. The stethoscope remains one of the most sought-after devices used for auscultation, referring to the art of listening to the heart and lungs' sounds in order to diagnose diseases related to the circulatory, respiratory, and gastrointestinal systems [1–5]. Auscultation simply involves auditing the heart and lungs' sounds for medical information.

Modern Practical Healthcare Issues in Biomedical Instrumentation
https://doi.org/10.1016/B978-0-323-85413-9.00007-4

Accurate and efficient auscultation may be challenged by low-frequency sounds produced by the heart, a noise-polluted environment, and noise-causing movement of the stethoscope chest piece. Hence, the user requires a good sense of hearing because of the sharp pace at which the heart performs its activities in order to avoid diagnostic error from lack of or inadequate concentration. An electronic stethoscope can enhance the sounds received through the head of the stethoscope and through a visual display to observe such sounds. The use of an electronic stethoscope has many benefits over the conventional one, and these could include improved frequency range, reduction of ambient noise, recording and replay capabilities and decreased time required for accurate reading, etc. By amplifying the sounds below the threshold, and also by using special filters that restrict the bandwidth by filtering out the frequencies that are not within the required range, the modern electronic stethoscope can resolve low sound levels. The introduction of electronic stethoscopes into the medical industry would make it possible to increase efficiencies that also allow the recording, storage, and reproduction of vital sounds [6–9]. For signal sampling, Arduinos can be used. Phonocardiography can be implemented using software engineering tools such as MATLAB or Python, which detail the graphical user interface (GUI) analysis of the audio signal. Biomedical engineering tools are common considering technological advancement, thereby allowing medical personnel to carry out auscultation with ease and little or insignificant errors. The electronic stethoscope is an engineering solution to this effect. The next section details the methodology with additional coverage of tests concerning the verification of the design and desired functionality to follow.

12.2 Methodology

A high-quality chest piece that can be used to handle body sounds from the skin surface to be fed into a microphone sensor for the conversion of electrical signals is included in this electronic stethoscope (or chest scope) device design. The MEMS (microelectromechanical system) sensor enables preamplification to increase the signal amplitude while shielding the sensor casing from external interference. Moreover, signal filtration eliminates noise, and constant amplification enables the sounds to achieve an audible level that can be clearly heard by the speaker. The signal is processed and digitized by the signal conditioning circuit, and further analyzed by an Arduino Due and MATLAB. A phonocardiogram is obtained in real-time with the measured heart rate in beats per minute [10–12], which is a representation of the heart sound. To obtain time and frequency analysis of the sampled and digitized data, the Fourier transform technique is used. Due to the features of the framework, this scheme can be used for training and clinical purposes.

Any physician can learn to use the electronic stethoscope with recurring exposure and to appreciate the process by which sound is obtained. An experienced physician appreciates the functions of the system such as serial data output, wireless transmission, and audio clip recordings due to sounds being transmitted electronically. The workflow of the proposed novel system is depicted in Fig. 12.1.

An advanced circuit that is capable of recording low-frequency heart sounds while filtering the unwanted frequencies for noise reduction is considered when constructing an electronic stethoscope, amplifying signals to a satisfactory level and then listening to the sound through a speaker. A small

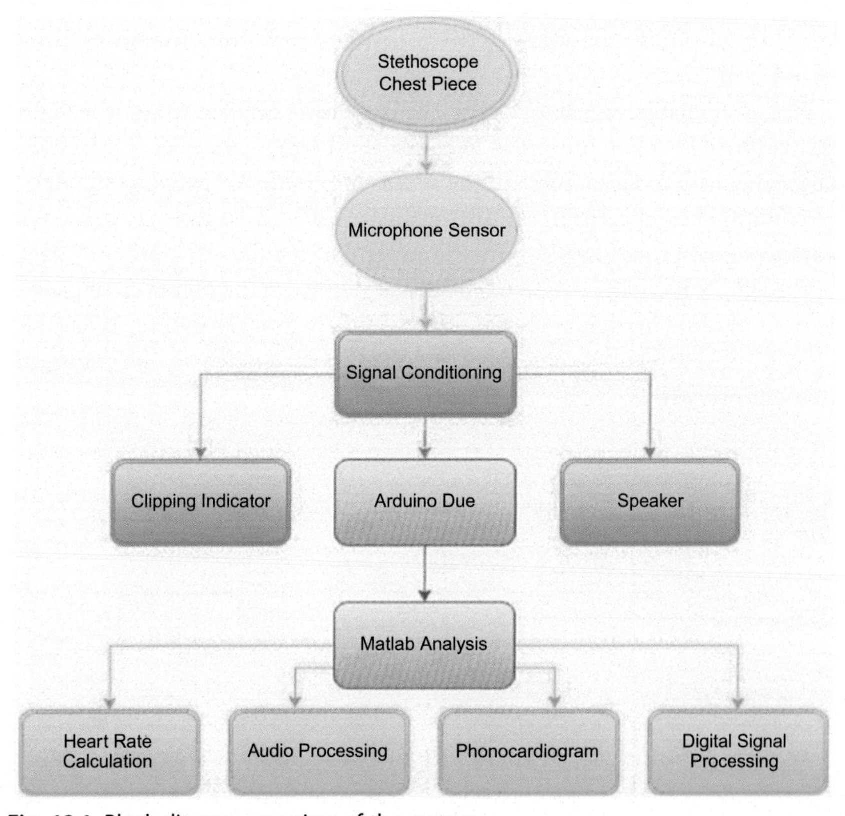

Fig. 12.1 Block diagram overview of the system.

sensitive microphone sensor that is used as the input device captures the heart or lung sounds that can be as low as 10 Hz; then, the acquired signal passes through a conditioning phase such as filtering and amplification, during which sampling and digitization are carried out before transmission for analysis to a central station.

A graphical depiction of the heart and lungs' sounds over time is a phonocardiogram. The heart rate was measured both in real-time and nonreal-time by detecting the dominant peaks. By transformation from the time domain to the frequency domain, the Fast Fourier Transform (FFT) provides the frequency elements of the time signal of the heart sounds.

To build the circuit, Proteus software was used. Before the printed circuit board design layout was made, the circuit was designed and examined on a breadboard. A three-pin head for connecting the microphone sensor used to record heart sounds is included in this design. The chest piece is attached to the 4th order Butterworth band-pass filter with the microphone sensor, which is a combination of a high-pass filter and a low-pass filter. The filter is based on the topology of the Sallen key with the op-amp used by LM741 [13].

Filtration enables signals of interest to be detected by limiting the signal's bandwidth. The filter output is fed into the power amplifier TDA2003, which gives rise to an increase in the signal amplitude for proper and effective audio accommodation. It also offers a DC offset that is useful for Arduino signal processing. A potentiometer that regulates the circuit's gain is included. As it could be used to check the validity of the component package used, the 3D modeling tool in Proteus was extremely useful. As shown in Fig. 12.2, several test points were included on the circuit to ease the measurement process.

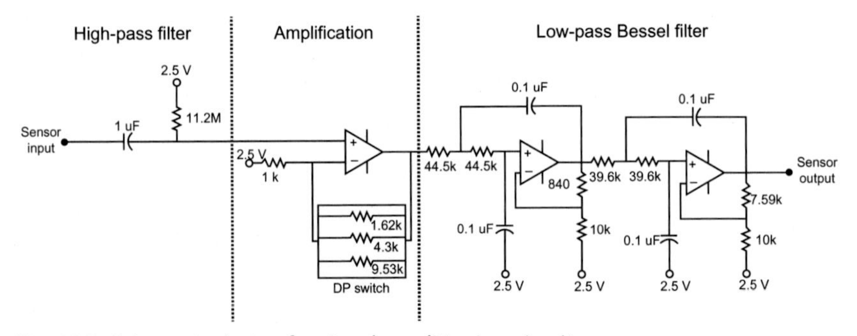

Fig. 12.2 Schematic design for signal conditioning circuit.

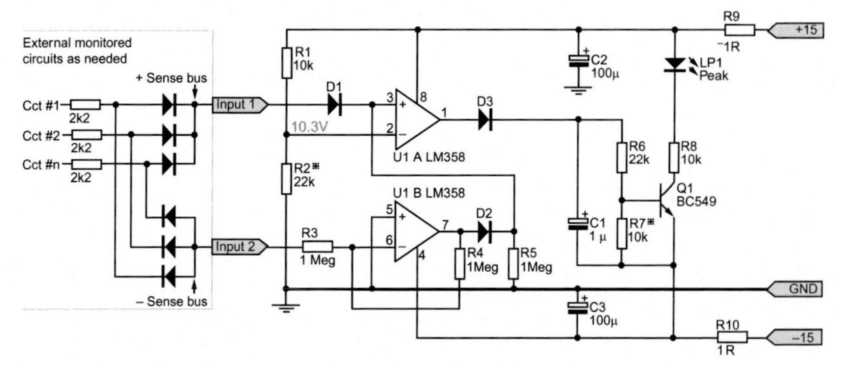

Fig. 12.3 Schematic design for clipping indicator circuit.

The audio amplifier output is sent to another circuit that tracks the amplifier's gain and warns whether clipping is occurring or not through a light-emitting diode (LED). To sample and digitize the raw analog data that was then used for further research, an Arduino Due was used. The circuit's highest frequency of interest is 150 Hz, so 500 Hz was chosen as the sampling frequency, equivalent to a sampling time of 2 min, taking into account the Nyquist theorem. The CoolTerm software was used to collect the information into a text file that was then analyzed and further processed using MATLAB, and using audio processing techniques, the text file was converted to wav. The architecture of the clipping indicator circuit is demonstrated in Fig. 12.3.

12.3 Tests, results, and discussion

Testing includes the software and hardware portion in which several tests have been carried out to validate the design of the computer, although other tests have also been conducted to fix system anomalies or faults. To obtain the time frequency features using the Fourier Transform, the data from the text file was analyzed in MATLAB. The MATLAB code was designed to read data from the microcontroller through a serial connection with a band rate of 9600 bps. The connection port was timely queried and read, followed by the processing and displaying on MATLAB in real time. A significant number of the strongest frequency harmonics from the Fourier Transform was analyzed and included in the phase spectrum in Figs. 12.4 and 12.5. Though harmonics have small similarity to the heart sounds, they do not originally sound like the heart. In this case, the heart sound and the main noise from the power supply were present and

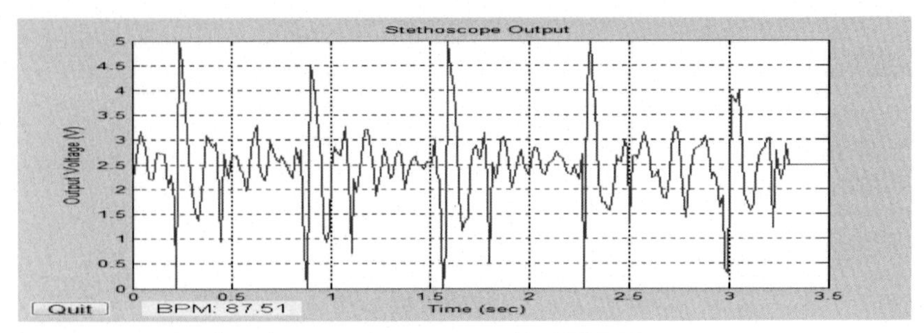

Fig. 12.4 Sample of cardiac waveform captured in MATLAB.

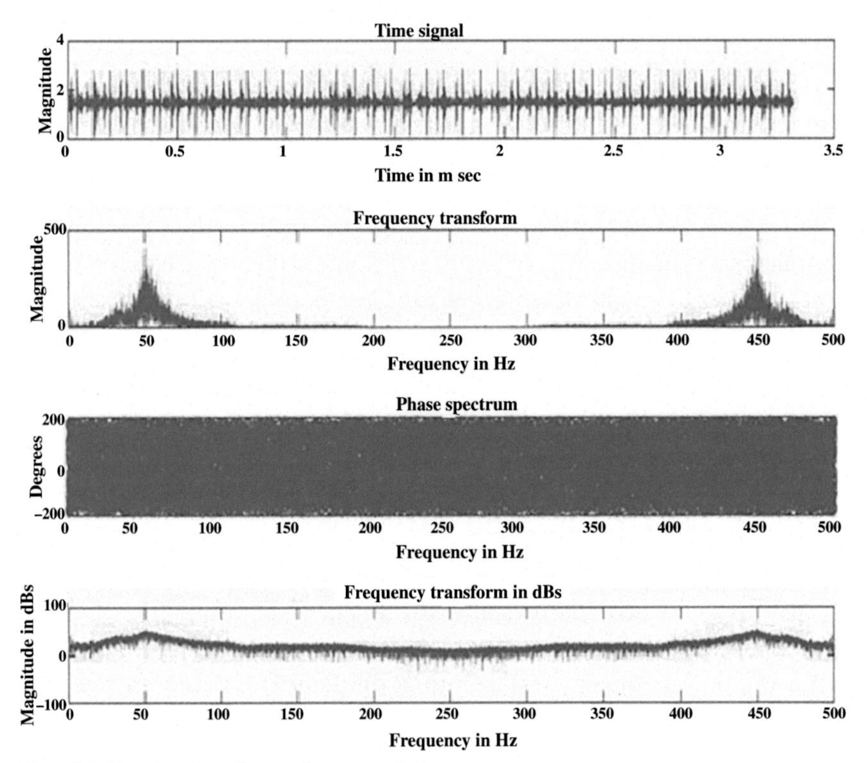

Fig. 12.5 Fourier Transform of captured data.

amplified, which makes them inseparable since they are both uncorrelated (Table 12.1, Figs. 12.6 and 12.7).

With a signal generator as the input to the circuit, the frequency characteristics of the filter were examined. A 20 mV signal was applied to the input, and the frequency ranged from 20 to 1000 Hz to observe the

Table 12.1 Calculated frequency response values.

Input freq. (Hz)	Gain (dB)	Input p-p (mV)	Output p-p (mV)
20	40.00	20	2000
50	41.94	20	2500
100	41.94	20	2500
150	39.78	20	1950
200	35.56	20	1200
250	32.04	20	800
300	28.79	20	550
400	23.52	20	300
500	19.08	20	180
1000	6.02	20	40

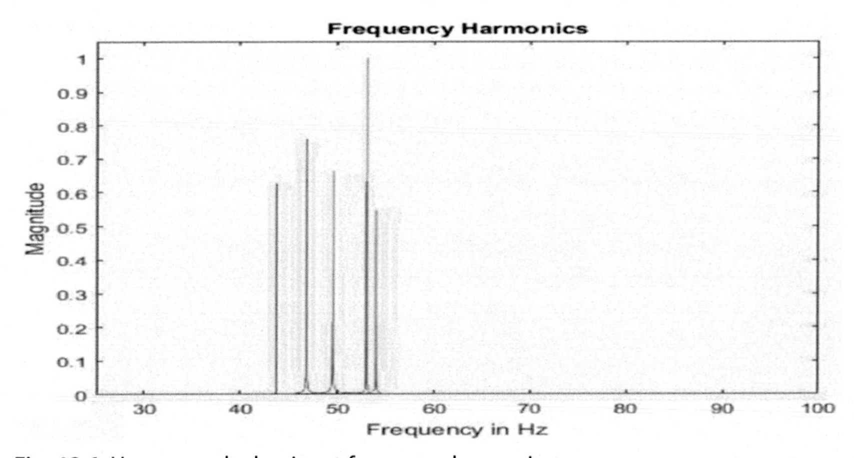

Fig. 12.6 Heart sounds dominant frequency harmonics.

amplitude of the signal at the frequency cut-off points. Inside the bandwidth, there is a significant gain that is shown to be the largest increase, and the data shows an estimated 3 dB decrease at the 150 Hz frequency cut-off. This means that the higher frequency signals were correctly reduced by the implementation of the 4th order Butterworth band-pass filter. The peak-to-peak of the input and output data (amplitude) means that the signal has been enhanced.

The input to the sensor circuit and the output of the audio amplification circuit were calculated using an oscilloscope. The performance in Fig. 12.8 (green (dark gray in print version) waveform@2.00 V/div) reflects the amplification of the input microphone's low-frequency sounds (yellow (light gray in print version) waveform@200 mV/div). The peaks correspond

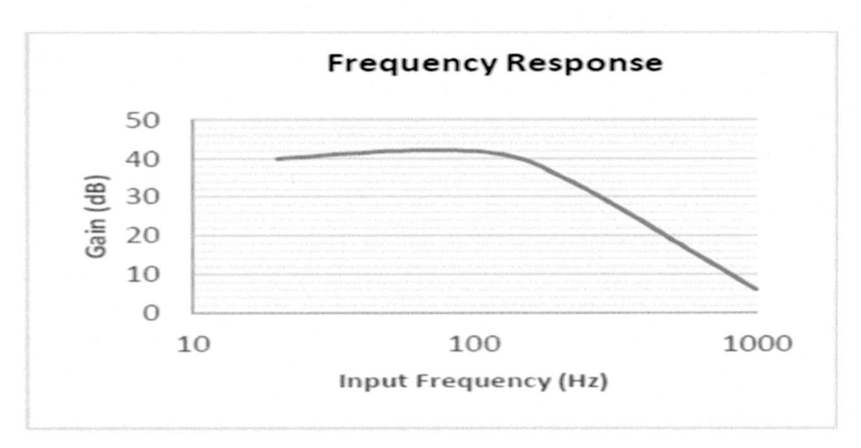

Fig. 12.7 Frequency response of analog circuit.

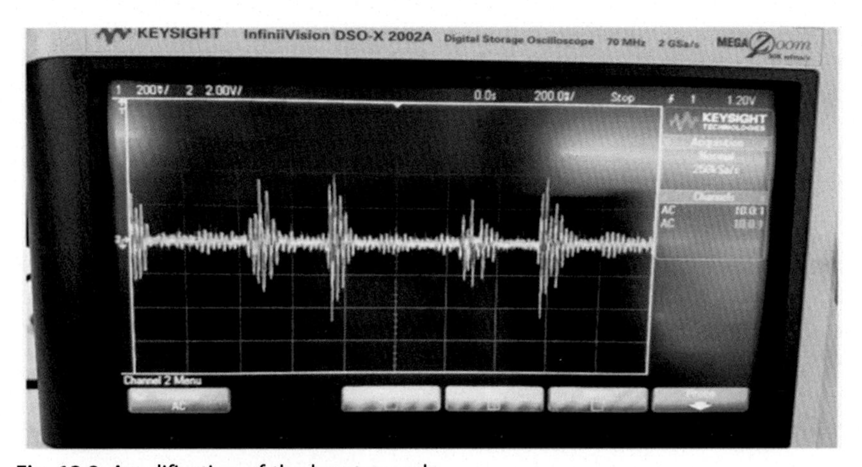

Fig. 12.8 Amplification of the heart sounds.

to the heartbeat that reflects the sounds of S1 and S2 and are also known as lub-dub. Due to the relaxation of the heart (diastole) and the contraction of the heart (systole), the signal between the heart sounds is a stage in the cardiac cycle, also known as the series of events occurring with each heartbeat [14–16].

From the lungs' sound, as shown in Fig. 12.9, there is an increase in amplitude during respiration and a considerable decrement in the amplitude during exhalation. Lungs' sounds are heard since most power from the lung sounds are within the same frequency range as the heart sounds (Fig. 12.10).

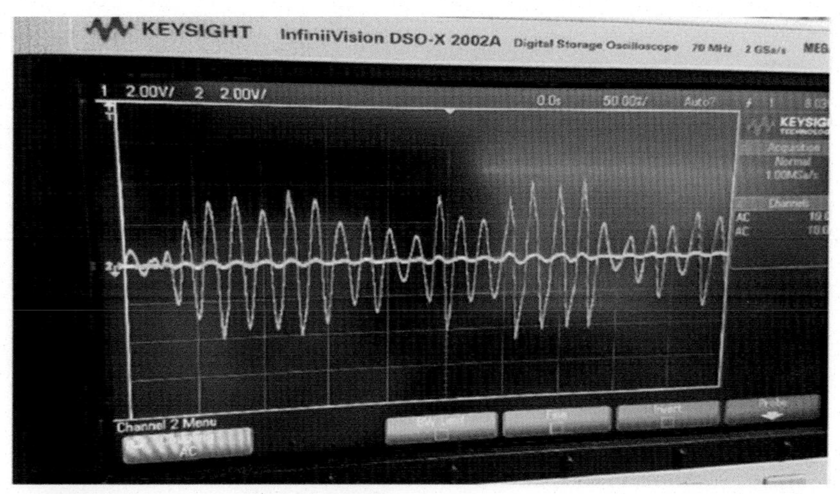

Fig. 12.9 Lung sounds during inspiration.

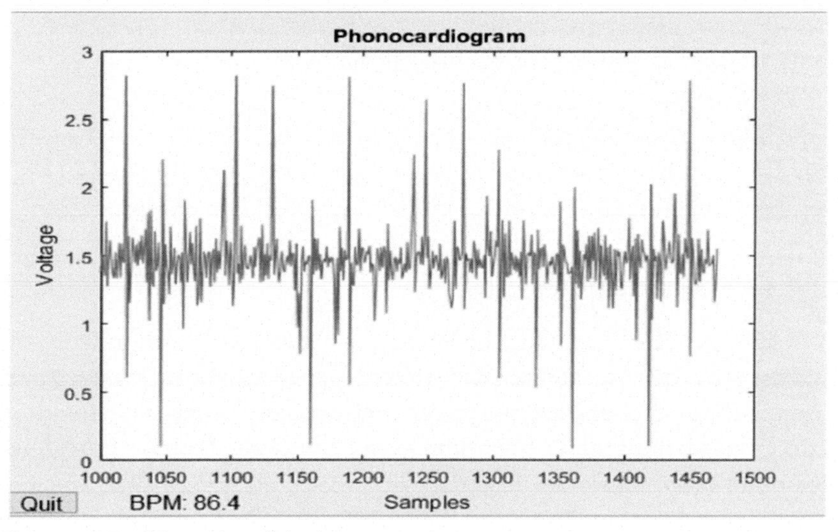

Fig. 12.10 Real-time plot of the heart sounds.

With the use of MATLAB to communicate with Arduino, the phonocardiogram data was plotted in real-time. MATLAB was able to read the voltage level on an Arduino analog pin and then plot the data [17]. The heart rate was determined by the average result extracted from the dominant peaks that existed in the data, thus providing the beats per minute (BPM) reading

Fig. 12.11 Hardware system of an electronic stethoscope.

of the hearts. The BPM readings are shown with the simple exit of the software through the user interface controls that are included in Fig. 12.11.

12.4 Conclusion

Conclusively, the system functioned perfectly and better (in terms of response time) than the conventional stethoscope and conforms to the specification of an original stethoscope that provides medical personnel with a perfect tool to hear subaudible sounds and murmurs, thereby providing opportunities for improved diagnosis. The system consists of two components: the hardware and the software components. The hardware primarily functions for the collection or acquisition of data while the software is used for postprocessing of acquired data for analytical purposes. This system is capable of filtering background noise and amplifying sounds from the heart and lungs so that the loudspeaker could make the sounds audible. In order to obtain the characteristics of the signal, the data was correctly sampled, captured, and further analyzed, and the phonocardiogram was shown alongside the heart rate being determined. The drawbacks of this system vary from the decrease in flexibility when taking measurements due to the attachment of the sensor with wires to the circuit and the inability to separate the noise from the signal from the power source.

A battery could be used to power the circuits to further boost the system, which would provide a fully standalone system to be used on the go. In order to relay signals wirelessly from the sensor to the signal conditioning circuit, a wireless transceiver, such as Bluetooth or Zigbee, may be integrated into the design, allowing patients to be examined at a distance [18]. In contrast to computer software such as MATLAB, creating a smartphone app would also enable data to be viewed wirelessly on a mobile device.

References

[1] M. Ahmed, V. Turnock, The history of the stethoscope, 2017, Retrieved from https://hekint.org/2017/01/27/the-history-of-the-stethoscope/. (Accessed 18 December 2020).

[2] H. Markel, The stethoscope and the art of listening, J. Med. 354 (2006) 551–553.

[3] D. Birrenkott, B. Wendorff, J. Ness, C. Durante, Heart & Breath Sounds Amplifier, University of Wisconsin-Madison Department of Biomedical Engineering, Madison, 2001.

[4] M. Grenier, K. Gagnon, J.J. Genest, J. Durand, L. Durand, Clinical comparison of acoustic and electronic stethoscopes and design of a new electronic stethoscope, Am. J. Cardiol. (1998) 653–656.

[5] S. Leng, R.S. Tan, K.T.C. Chai, et al., The electronic stethoscope, Biomed. Eng. Online 14 (2015) 66, https://doi.org/10.1186/s12938-015-0056-y.

[6] C. Ahlstrom, Processing of the Phonocardiographic Singal-Methods for the Intelligent Stethoscope, *Department of Biomedical Engineering Linkoping University Institute of Technology*, 2006, p. 32.

[7] W. Wah, D. Myint, D. Bill, An electronic stethoscope with diagnosis capability, in: *Proceedings of the 33rd Southeastern Symposium on System Theory*, 2001, pp. 133–137.

[8] M. Wu, G. Der-Khachadourian, Digital Stethoscope—ECE 4760, [online] Available: https://people.ece.cornell.edu/land/courses/ece4760/FinalProjects/s2012/myw9_gdd9/myw9_gdd9/.

[9] R. Priya, A. Cheeran, V.D. Awandekar, R.S. Mane, Remote monitoring of heart sounds in real-time, Int. J. Eng. Res. Appl. *3* (2) (2013) 311–316.

[10] M. Gogoi, An user friendly electronic stethoscope for heart rate monitoring, J. Appl. Fundam. Sci. 1 (2) (2015) 233.

[11] T. Chakrabarti, S. Saha, S. Roy, I. Chel, Phonocardiogram signal analysis—practices trends and challenges: a critical review, in: *2015 International Conference and Workshop on Computing and Communication (IEMCON)*, 2015.

[12] M. Santos, M. Souza, Detection of First and Second Cardiac Sounds Based on Time Frequency Analysis, Electronic Department E.E. Federal University of Rio de Janeiro, 2001.

[13] M.S. Obaidat, Phonocardiogram signal analysis: techniques and performances, J. Med. Eng. Technol. (1993) 221–227.

[14] V.K. Ingle, J.G. Proakis, Digital Signal Processing Using Matlab V.4, PWS Publishing Company, 2016, pp. 1998–1999.

[15] K. Courtemanche, V. Millette, B. Natalie, Heart Sound Segmentation based on Mel-Scaled Wavelet Transform, Faculty of Science McGill University, 2009.

[16] A. Atbi, S.M. Debbal, F. Meziani, Heart sounds and heart murmurs sepataion, in: *IWBBIO* 2013. *Proceedings*, 2013, pp. 265–273.
[17] Arduino Support from MATLAB— Hardware Support—MATLAB & Simulink, Feb 2017. [online] Available: https://uk.mathworks.com/hardware-support/arduino-matlab.html.
[18] Y. Luo, Portable bluetooth visual electrical stethoscope, in: *Communication Technology (ICCT) 2008 IEEE International Conference*, 2008.

CHAPTER THIRTEEN

Application and impact of phototherapy on infants

Dilber Uzun Ozsahin[a,b,c], John Bush Idoko[d], Nyasha T. Muriritirwa[b], Sabareela Moro[b], and Ilker Ozsahin[a,b,e]

[a]DESAM Research Institute, Near East University, Nicosia, Turkish Republic of Northern Cyprus, Turkey
[b]Department of Biomedical Engineering, Near East University, Nicosia, Turkish Republic of Northern Cyprus, Turkey
[c]Medical Diagnostic Imaging Department, College of Health Sciences, University of Sharjah, Sharjah, United Arab Emirates
[d]Applied Artificial Intelligence Research Centre, Department of Computer Engineering, Near East University, Nicosia, Turkish Republic of Northern Cyprus, Turkey
[e]Brain Health Imaging Institute, Department of Radiology, Weill Cornell Medicine, New York, NY, United States

13.1 Introduction

Jaundice is a medical condition that affects newborn babies, causing the skin and whites of the eyes to turn yellow. Jaundice in premature children is referred to as neonatal jaundice. Neonatal jaundice is one of the most common disorders affecting newborn babies, with 60% of babies expected to experience jaundice soon after birth, including around 80% of pregnancies less than 37 weeks.

Bilirubin is the substance that causes the yellow color in the blood, and 1 in 20 children have blood bilirubin levels high enough to need treatment; it's quite manageable whether detected soon after birth or 6–48 h afterwards, possibly even longer. This is a very common and usually harmless condition for newborns. Though in some number of cases it does present considerable difficulty in managing it when very low birth weight babies and/or premature babies with low birth weight are jaundiced. Because most of these examples would be difficult cases, these babies would be plagued by serious illnesses like apnea of prematurity, sepsis, respiratory distress syndrome, and some other underlying health conditions. In this small number of cases, jaundice will be a sign of the anomalous health conditions underlying the newborn. If tests indicate a very high level of bilirubin in the blood of an infant, treatment for newborn jaundice is recommended as there is a substantial risk that bilirubin will move into the brain and cause brain damage. Therefore, to quickly minimize the treatment of the baby's bilirubin levels,

Modern Practical Healthcare Issues in Biomedical Instrumentation
https://doi.org/10.1016/B978-0-323-85413-9.00011-6

there are two key treatment methods that can be done in the hospital: phototherapy and exchange transfusion. Neonatal jaundice phototherapy would avoid the increase of total serum bilirubin to dangerous levels and catalyze the immediate breakdown of the serum in combination with the baby's system. This would be a better option than the exchange transfusion. Phototherapy can minimize the peak total serum bilirubin and jaundice length and remove the need for an exchange transfusion, thus avoiding the total occurrence of neurological problems associated with bilirubin in the infant. The peak total serum bilirubin level would be effectively decreased by phototherapy.

13.1.1 Newborn jaundice

Jaundice, medically known as neonatal jaundice in babies, is quite a common occurrence in newborn babies with the primary stimulating cause being too much bilirubin in the baby's blood (hyperbilirubinemia). Bilirubin is a yellow substance that is produced when red blood cells are broken down in the blood stream, and its direct pathway is through the blood stream to the liver.

Soon after the baby is born, particularly within the first few hours after birth, the liver is not entirely developed and ready to break down the large amounts of bilirubin that will be forming in the blood rapidly because the baby's physiology is no longer linked and dependant on the mother's for all physiological functions and processes. Thus this results in the levels of bilirubin rising consequently as the liver cannot keep up with respect to bilirubin production and because the liver will not be operating at optimum efficiency to breakdown and remove the dangerously increasing bilirubin quantities in the baby's blood. Health institution officials in midwifery and infant care have a visual and analytical system to detect and determine therapy for all newly born babies if they are or are not jaundiced within the first few hours after birth as to maximize treatment procedure effectiveness. See Table 13.1 for management of hyperbilirubinemia. Recent studies are giving data to suggest that even healthy term infants may suffer mild neurologic damage with bilirubin concentrations >20 mg/dL [1].

Apart from cases where the baby's liver is underdeveloped significantly for the task, there are various attributes for hyperbilirubinemia and we can attest to some deficient uptake, close to cut-off amounts of ligandin, augmented resorption of bilirubin, and a severe decreased capacity to secrete cholephilic substances/compounds as well.

Table 13.1 Tabulation of neonatal bilirubin concentration within the first 72 h after birth, four-stage over watch monitoring with treatment options at each stage for neonatal care.

		Treatment	
Age (h)	**Bilirubin (mg/dL)**	**Phototherapy**	**Exchange transfusion**
<24	Visible jaundice	Consult attending physician	
25–48	>15	×	
	>20	×	×
49–72	>18	×	
	>25	×	×
>72	>20	×	
	>25*	×	×

Neonatal jaundice also stems from transient deficiency conjugation, combined with an exponential increase in the turnover of red blood cells. There are a number of pathological conditions that shoot up bilirubin production, and some are isoimmunization, heritable hemolytic disorders, and also blood extravasations, which come from bruises and cephalohematomas. The factors that lead up to the possibility of neonatal jaundice are inherently different, for the greater percentage of infants that end up in need of treatment are attributed to the common causes and path epidemiology. Late-preterm babies and those solely breast-fed are the largest group of otherwise stable infants at elevated risk of hyperbilirubinemia (particularly if breastfeeding is not going well). Breast-feeding and low caloric intake associated with problems with breastfeeding are both thought to cause an increase in bilirubin enterohepatic circulation [2].

The total red blood cell death count, in relation to bilirubin breakdown, will be at minimal levels in the first 6–9 h after birth with a steep upward slope as total bilirubin concentrations augment without an effective breakdown rate or compensated reaction that must be counteracting the steady and hazardous levels of bilirubin production. Liver immaturity and metabolism tardiness for incomplete child development factor into an increased risk for total red blood cell death and bilirubin production (mg/dL) against discrete time intervals. If these factors are not rectified immediately, they will give rise to numerous complications as a result and impairment in the developing infant. Definitive high-risk patients commence therapy without delay as those monitored ranges would have likely been surpassed. Intermediate-risk patients are retried for therapy, and some are reviewed depending on hepatic and blood conjugation factors along with future risk if therapy need is due primarily within a preinformed time interval for recovery and self-correction. Some commence therapy though estimates and data would classify them as not high-risk. For low-risk patients, repeated checks and tests are cautious confirmation for those with potentially jaundice-inducing factors, but if they are in the safe zone, most of them will be on standby unless visible symptoms show but, otherwise, are deemed sufficient for self-deliberation.

Throughout the immediate hours after birth, all risk-level patients are closely monitored with help and therapy ideally identified and applied; risk exposure is high for prematurely born babies, and definitive curve monitoring is requested soon after birth with a viable physician-approved treatment plan.

In most cases, this will be an overall data chart to note statistical assessment of how many newborns for a particular day eventually became jaundiced and their conjugate bilirubin concentration over their initial and most vulnerable days after birth.

Also, babies whose blood type is not compatible with that of their mothers can develop a build-up antibody that can destroy their red blood cells and cause a sudden rise in bilirubin levels. Unfortunate cases where there are liver problems, possible infection, enzyme deficiency, and abnormalities in the baby's red blood cells will lead to a mass death of erythrocytes with an augmented response of bilirubin level increase. Such reactionary outcomes are either genetic or causal because these high levels of bilirubin in the blood stream in turn lead to numerous complications in babies. Due to the developing central nervous system components that require efficient blood flow, which will be contaminated with very high levels of bilirubin, this will lead to bilirubin encephalopathy, consequential kernicterus, and permanent neurodevelopment disabilities.

13.1.2 Classification of neonatal jaundice

(a) Conjugated jaundice

Physiological jaundice is often identified in 86.5% of full-term newborns. Transient hyperbilirubinemia is caused by rapid hemolysis of the newborn, as temporary immaturity enzymes in the liver. Some classification will be under unconjugated hyperbilirubinemia. Without any early intervention, the general condition of the newborn baby results in yellowness of the skin that appears 2–3 days after birth and disappears at 10–14 days. Bilirubin will rise at a rate of less than 5 mg/dL, having peak quantity saturation in the blood stream between 3 and 5 days, with a limit that will never exceed 15 mg/dL. Similar conditions will be prolonged in prematurely born babies with an approximate duration of 4–6 months of conditional awareness in the baby's system. Due to hepatomegaly, the liver will enlarge slowly, from 3.5 cm below the right costal, and this will give rise to most expectant splenomegaly from the extrahepatic portal venous obstruction. If there is another underlying similar condition within the newborn, it will be most likely absent; feces and urine will be unpainted, signaling no fetomaternal hemorrhage or fetus anemization.

(b) Hemolytic jaundice

This type is characterized as hemolytic disease of the newborn and microspherotic congenital hemolytic jaundice due to congenital enzymopathies. Crigler-Najjar Syndrome (CNS) is a rare enzymopathy caused by congenital deficiency of the enzyme, glucuronyl transferase, in the liver cells, leading to a disruption of binding and releasing of bilirubin. This type of clinical jaundice, which occurs at birth, continues throughout the child's life; the amount of indirect bilirubin increases from approximately 256 to 765 mmol/L. There is a crucial need for symptomatic treatment for this particular jaundice type, and even in severe cases, greater care is taken for aggressive treatment with most of the children affected dying before their first birthday due to CNS. Death rate by CNS associated randomly with maternal hemolytic anemia, which has probability of 1–3 death per 100,000 newborn per year and detection is early in order to search for underlying systemic autoimmune disease that may lead to Gilbert syndrome. Dubin-Johnson Syndrome is also a rare transient potential outcome causing isolated augments of conjugated bilirubin in serum. Autosomal recessive Rotor type hyperbilirubinemia indirectly causes a combination of direct-unconjugated and indirect-conjugated hyperbilirubinemia, though it'll be nonhemolytic jaundice due to chronic elevation of predominantly conjugated bilirubin.

(c) Mechanical jaundice

Mechanical jaundice is caused by conditions that prevent bile from flowing normally from the liver to the intestines. Such conditions include: biliary stricture (narrowing of the bile duct) cancer of the pancreas or gallbladder. Specialists may perform the following tests to determine the cause of mechanical jaundice: magnetic resonance imaging, computed tomography scanning, endoscopic retrograde cholangiopancreatography, and bilirubin levels examined by blood tests.

13.1.3 Treatment of jaundice

The treatment to alleviate newborn jaundice is varied and determined by the severity and or nature of the cause that led the baby to be jaundiced. The main objective that will be achieved by therapy is to lower the concentration of circulating bilirubin and/or prevent it from building up. With decreased bilirubin levels, normal infant development will progress. Neonatal jaundice treatment strategies:

1. Exchange transfusion
2. Phototherapy
3. Intravenous immunoglobulin G (IVIG) therapy

These schemes are implemented carefully after determining the nature of the cause of the hyperbilirubinemia. An absolute blood exchange for disease-natured causes with an older individual for bilirubin process through metabolic breakdown by a developed liver results in downsizing the bilirubin concentration in the bloodstream. An exchange transfusion is a procedure that is done to counteract the effects of serious jaundice; the baby will receive small amounts of blood from a donor that replaces the baby's damaged blood with healthy red blood cells. This also increases the baby's red blood cell count and reduces bilirubin levels, and in very severe cases, the procedure may require that the baby's blood be removed and passed through somebody else's circulatory system so that the bilirubin is processed by that person's system, thereby lowering the overall amount of bilirubin in the baby's bloodstream.

Intravenous immunoglobulin G (IVIG) therapy is used for Rh-hemolytic disease hyperbilirubinemia to minimize the need for an exchange transfusion for the infant. An exchange transfusion is definitely needed for any baby diagnosed with hyperbilirubinemia with a count rising by 8.5 μmol/L per h (0.5 mg/dL per h) or more. However, administration of IVIG with severe hyperbilirubinemia, which is injected into a vein for an hour or more, eliminates the need for an exchange transfusion without causing immediate adverse reactions.

Phototherapy is the widely used therapeutic method for treating all jaundiced newborns. It is also the most effective treatment when particular acute and efficient wavelengths are employed. Optimal wavelengths of light should be applied in order to get results greater than 25% photo isomer formation, particularly along the blue spectrum (390–470 nm). Phototherapy does this specifically by using light energy to modify the form and structure of bilirubin, such that no undesirable effects are encountered by the infant from the infrared and ultraviolet ranges, transforming it into to compounds that can be expelled although normal conjugation is clearly deficient. Via the absorption by the upper dermal and subcutaneous bilirubin of the blue light used, a fraction of the pigment is induced to undergo many photochemical reactions that occur at different rates relative to each other. The yellow stereo isomers of bilirubin and some colorless derivatives with a small molecular weight are provided by such processes. The products are less lipophilic than bilirubin and can be excreted in bile or urine without the need for conjugation, unlike bilirubin. The relative contributions of these various reactants to the overall removal of bilirubin are still unknown, although studies in vitro and in vivo indicate that photo isomerization is more significant than image degradation [3].

Since it is clearly shown that bilirubin absorbs light primarily in the blue region of the visible light spectrum near 460 nm, the penetration of skin tissue by light increases further with increasing wavelengths. The rate of formation of photoproducts of bilirubin is now heavily dependent on the intensity and wavelengths of light in use, so bilirubin can only absorb unique wavelengths that can pass through the tissue, resulting in a phototherapeutic effect. For the treatment of hyperbilirubinemia, LEDs or lamps with outputs mainly in the 460–490 nm blue area of the visible light spectrum, are probably the most effective [4].

13.1.4 Effectiveness of phototherapy for neonatal jaundice treatment

Application of phototherapy for neonatal jaundice treatment has been mostly effective since its development and usage. Systematic analysis of the effectiveness and safety regarding different wavelengths employed; varying intensities used; total dosages given to the infant, whether overall or just discretely; and the defining threshold for commencement of therapy as recommended by an attending physician, hospital, or health official will paint a clear picture. It is generally accepted that continual phototherapy applied to a diagnosed infant will greatly reduce the high serum bilirubin within their bloodstream, inherently reducing, and in most cases removing, the need entirely for exchange transfusion in infants diagnosed with hemolysis.

Compared to traditional phototherapy, the disparity between blue LED lamps is much less important in reducing mortality in preterm infants needing phototherapy. Double conventional phototherapy has been largely more effective than single conventional phototherapy in decreasing the length of treatment and the mean serum bilirubin level with hemolysis included in terms of infant birth weight 2500 g or above. It is still not known, however, whether double phototherapy decreases the need for an exchange transfusion. It is not yet understood if even triple phototherapy improves efficiency. It is not clear if there is any efficacy in intermittent hospital phototherapy with threshold hospital phototherapy in comparison with light in the different total doses used for hospital phototherapy for unconjugated hyperbilirubinaemia [5–9]. It is widely agreed that only if serum bilirubin levels exceed predefined thresholds can phototherapy be applied.

As shown in Table 13.2, the several factors that have been mentioned previously have overall responsibility to the physiology and general development of the newborn with the infant's genetic history becoming crucial in assisting certain negative developments from phototherapy sessions or, in some cases, helping the infant to withstand any of these side effects that may even surface. Thus for most babies there are basic normal effects that are standardized, noted as normal after procedure, and seemingly disappear and reappear within specified periods; then some consequent circumstances surface in conjunction with other internal factors of the infant. Unfortunately, some are known to result in irreversible conditions that are rarely fatal (Table 13.2).

The duration of the therapy and what the blue wavelength light does to the infant's system are mostly accounted for before commencing therapy when risk minimization routines are advised and the attending physician recommends a viable plan that should be adequate for the infant's health and recovery with minimal to no effects otherwise. In most scenarios, infant reactions are normal and can be accounted for with ready remedies in place. Though some circumstances, like the time lost between the mother and the

Table 13.2 Prejudicial interests arising from neonatal phototherapy.
Neonatal phototherapy prejudicial precipitates

Short-term	Long-term
Thermal regulation instability	Escalation of development of child asthma with a high occurrence percentile 140%
Mother-baby bonding affected	Elevated risk for type diabetes mellitus, 379% percentile
Gastrointestinal tract movement abnormalities	Potential skin cancer for infant undergoing therapy
Baby flux and dehydration	Allergic diseases early development
Bronze baby syndrome	Melanocytic nevi
Electrolyte disturbance	Retinal damage
Circadian rhythm disorder	Patent ductus arteriosus

infant to bond, cannot be replaced, and resulting infant dehydration will just have to be risked. Table 13.2 depicts the analysis of the therapy for the babies undergoing continued phototherapy sessions.

13.2 Phototherapy

There are basic orientations for employing phototherapy and are all dependant on whether the patients are in a facility or outside official health premises. Exposure to fluorescent light bulbs or other light sources, such as halogen lamps, LEDs, and even sunshine, for treatment is often called light therapy or heliotherapy. This light treatment is the method of using light to remove bilirubin in the blood by placing an infant in the shine of light on their skin, and the blood will absorb these light waves that convert bilirubin into products that can move through their system [10–14]. The way this is accomplished is by applying monochromatic visible light in the fastest photo isomerization reaction attainable by high-pressure liquid chromatography from the 350 nm to the 550 nm scale. This was the most powerful in vitro wavelength, resulting in serum breakdown of more than 25% photo-isomer when therapy was being developed and verified to administer maximum photo isomerization in the blue spectrum from around the range of 390 to 470 nm. Green light (530 nm) is not only inadequate for the processing of photoisomers but is also capable of reversing the reaction in some situations, thus implying that the blue portion of the visible spectrum must be included in any clinically beneficial phototherapy unit, suggesting that the efficacy of phototherapy can be improved by the removal of green light.

The use of the various wavelengths is dependent on the therapy requirements with frequency increase corresponding to penetration power, reverse correspondence being less penetrative, and more cell effect seen [15–17].

Both factorial brilliance and blue shade are precisely calibrated for optimum pediatric therapy success, noting the lack of adequate irradiation brilliance trailing the process and, in some cases, being ineffectual entirely as noted in the <600 nm range with inadequate brilliance.

Different phototherapy devices, such as the conventional phototherapy light and the fiber-optic phototherapy device, are currently available on the market. Fluorescent or wide-emission spectrum tungsten halogen lamps or narrow-spectrum LEDs are the common light sources used in these devices. As a treatment plan for neonatal jaundice reduction, hospital phototherapy is administered either by standardized means or through fiber-optic lights. The use of light for hospital phototherapy with different wavelengths for unconjugated hyperbilirubinaemia in term and preterm infants has different interventions, including daylight fluorescent lamps, conventional phototherapy (using halogen-quartz bulbs), narrow spectral emission blue fluorescent lamps, standard blue fluorescent lamps, blue-green LED lamps, blue-green fluorescent lamps, blue LED lamps, and green fluorescent lamps. The more effective choice than basic blue fluorescent light could be from the variety of options for use of blue-green fluorescent light to reduce the requirement for prolonged phototherapy after 24h in most healthy low birth weight babies with hyperbilirubinaemia in the first 4 days of life. Compared to traditional phototherapy (using halogen-quartz bulbs) in term and preterm babies, many hospitals prescribe phototherapy using blue LED lamps that have been effective in reducing the amount of hours spent on phototherapy [18]. When used, the different wavelengths of light vary in their effects on the rate of decline in serum bilirubin levels. Compared to distant light-source phototherapy, near phototherapy can decrease the length of phototherapy and the mean serum bilirubin level in infants when treatment is applied. In general, hospital procedures do not require the initiation of therapy at such particular limits. In significantly low birthweight infants, lower thresholds compared with higher thresholds may reduce the proportion of infants with neurodevelopmental disability, profound impairment, and significant hearing loss. The common main two methods employed are fiber-optic phototherapy and phototherapy light.

Fiber-optic phototherapy: This is a customized and much more linear and fluid application of the method. The use of fiber optics streamlined the transfer and improved luminance with greater reduction in artifacts that come from conventional light-powered sources and not from either light-emitting diode projection and luminance transferred and preserved within wavelength range without superimposing certain wavelengths on progressive waves en route for illumination on the infant. Incorporating fiber optics gives us the benefit of making use of different light sources that are easier to coordinate and can be fabricated to required malleable specifications to fit our design needs. Fluorescent lights and LEDs both are incorporated into the devices with their smaller component design assimilating better in a widespread approach for transfer through fiber optics and projection of the same powerful light source that is produced from power supplied for the device preserved and delivered adequately.

Phototherapy light: This is the basic design and is the most common orientation in manufactured models. It makes use of the light source above the infant, connected to the normal room power supply, and is capable of being operational for more than 4 h with stops for feeding the infant and for mother-baby bonding times. This is the normal and common type used in many hospital maternity wards and IVF centers as they are for onsite treatment with the infant and mother present in the facility for the duration of the therapy. They are most often used due to their stationery nature and simple design featuring a baby cot compartment and illumination above with provisions for covering the infants' eyes during a session [19].

13.2.1 Duration of treatment

For most standardized concerns, such as the American Academy of Pediatrics guidelines, phototherapy is required to be needed for 24 h or less, and in some cases, it may be required for 5–7 days, even slightly longer in other cases. Babies with a bilirubin serum concentration of <18 mg/dL are not recommended to discontinue phototherapy treatment until serum levels have lowered in the 13–14 mg/dL range below the safe newborn bilirubin range. The use of adequate and responding light sources for the therapy has been known to influence treatment duration with blue fluorescent lamps compared with green fluorescent lamps: blue fluorescent lamps and green fluorescent lamps seeming equally effective at reducing the duration of phototherapy in both term and preterm infants with nonhemolytic jaundice [2]. For both blue and green lamps, we do not have adequate evidence-based data to conclusively state which light source gives the least amount of therapy for similar conditions if both colors are in use. Due to this lack of conclusive evidence, we do not have enough information to clearly state how blue fluorescents and green fluorescents particularly differ in effectiveness at catalysis for the rate of reduction in serum bilirubin for term and preterm infants, especially in low birthweight infants with nonhemolytic jaundice who would be stratified by an initial serum bilirubin level status of 9.0–12.0 mg/dL, 12.1–16.0 mg/dL, and/or 16.1–21.0 mg/dL. No evidence has been found from randomized controlled trials on the mortality factor for these infants. Although the following are within therapy thresholds: neurological/neurodevelopmental outcomes, the need for exchange transfusions after the treatment period should be noted and effectively used for therapy if necessary and recommended after serum level count.

13.2.2 Baby pouch design

The current conventional means of how phototherapeutic treatment is applied is considerably effective ever since its inception in medical research, and even after treatment application has been slightly modified, in the 1960s when trials were run to assess total effectiveness in bilirubin reduction, breakdown in upper and lower dermal layers, and overall impact on the baby's developing physiology as treatment is delivered over continuously long hours [6]. Current methods involve the baby under a light source projection in a crib designed for total light illumination on the skin for a long duration, light indefinitely shining, in interval of 4 h for baby feeding, diaper changes, and cuddle breaks; the light is continuously absorbed by the baby with eye patches for eye protection. Newer developments incorporate a blanket in which the baby is covered, as it is in essence a blanket just incorporated with LEDs all over its inner surface and fiber-optic lining bringing the light toward the blanket area. Some hospitals and private health institutions offer the blanket as a finer option for infants who might need a more delicate and hands-on treatment option, as the alternative procedure involves long hours where the baby's alone or asleep under the light source for the therapy session. The bili-blanket brought a smoother and kinder option for mothers to be able to be holding their newborn and breastfeed concurrently as the blanket is online and therapy is underway, though it is not mobile and is specialized to be indoors and operated by the designated nurse.

The proposed baby pouch will be easy to operate with basic straightforward controls for it to be used by virtually anyone in mobile contexts with an optional power source for mobile or stationary use. A polymer-based light transferable and translucent design in the shape of a pouch or baby jumpsuit [12] would make it wearable for the baby already wearing clothes, though it is optional, and could cover the baby from the neck to the feet as well as the upper extremities. The brightness will be able to encompass the whole baby if optional brightness is required when the baby is sleeping. This pouch design will be available for comfort for parents as it would allow them to continue normal day-to-day happenings and still be with their newborn as the baby also receives therapy.

The baby pouch will be comprised of the power section, polymer pouch, and display section for control input and timer visualization (see Fig. 13.1).

With acquiring all the components within the three sections of the baby pouch design, fabrication of the translucent polymer with LEDs and fiber optics for light movement to improve light output will be incorporated with adjustable brightness based on power input with variable power accounting for the incrementally similar brightness.

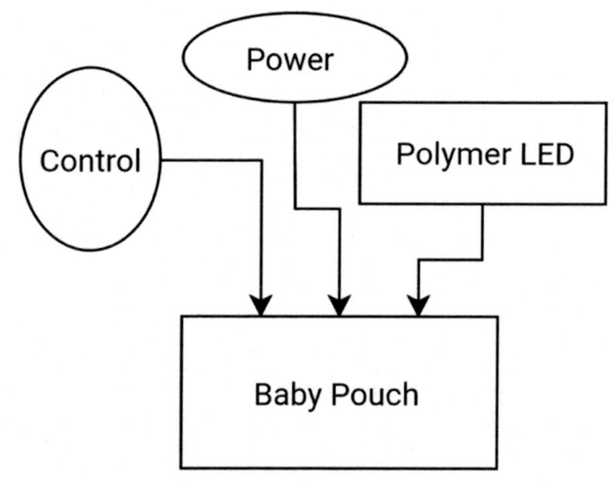

Fig. 13.1 Block component build-up of entire design.

13.2.3 Baby pouch design circuitry

The various electrical components that are required to compose this device are modestly reliable and acceptable to fabricate. Power regulator components include 2 (1×10^{-6} F) capacitors, 1 (100×10^{-6} F) capacitor, 2 diodes (1N4149), 1100-Ω resistor, 1100-Ω variable resistor, LM317 voltage regulator, LED display input/output, internal rechargeable battery, and plug-in. Polymer pouch component consist of fabricated polyethene pouch, hypersensitive blue-tuned LEDs, power microcontroller timer, pouch straps, and covering template (possibly polythene material) [3]. All available components will be premanufactured and simply acquired for assembly of the device. All safety and ISO standards are subsidized by use of limited- to low current–voltage output components, and this facilitates safe and regular family use more.

13.3 Conclusion

The main objective of developing the baby pouch for a phototherapy treatment procedure is to reduce procedural costs that are incurred in obtaining treatment by normal conventional means. For most hospitals and health institutions that offer phototherapy treatment for newborn jaundice, costs vary from $85–150 per day for each session of conventional therapy. Such will amount to at least $5000 for a whole month of therapy for effectively low bilirubin level patients. Treatment will always exceed the minimum average of 2 weeks, and if severe or if the baby has other complications, it will be incorporated with other treatments, thereby incurring more costs. With the baby pouch, once it is paid off, it will ensure a whole 2 months of effective home-based therapeutic treatment with just fortnight visits to the hospital for assessment and evaluation of bilirubin levels in the baby's physiology. It is advisable to ensure resale or a rental process of the pouch for continued effective use.

Treatment of jaundice with conventional phototherapy that requires daily sessions under the light source at the hospital and uneconomic treatment plans is only viable for those who can afford and have integrated medical aid plans. If phototherapy does not counteract the effects in the required time, more aggressive methods or alternative methods will be employed. With IVIG treatment being more costly per mol of the substance used in therapy, in most cases, total cost of IVIG therapy ranges from $5000 to $10,000, depending on the number of infusions per course; costs may include a hospital stay if home infusion is not covered with medical aid or insurance. With certain exceptions of severe cases with Rhesus disease and treatment alike, this compounds treatment expenses and, in most developing countries, is unlikely to be affordable for most families.

Therefore, it is evident newborn jaundice is treatable with a wide range of therapies that have been proven to be effective in eliminating it entirely. A reduction in costs, this being the main hiccup, are to be encouraged by development and use of the baby pouch. This product, once manufactured and applied, will assist many and is reasonably profitable for current markets for medical equipment supply sales and production costs. It could be a major economic frontier for all those who are not able to afford today's ever increasing child health care.

References

[1] A. Montealegre, N. Charpak, A. Parra, C. Devia, I. Coca, A.M. Bertolotto, Effectiveness and safety of two phototherapy devices for the humanised management of neonatal jaundice, An. Pediatr. (English Edition) 92 (2) (2020) 79–87.

[2] M. Namnabati, M. Mohammadizadeh, S. Sadari, The effect of home-based phototherapy on parental stress in mothers of infants with neonatal jaundice, J. Neonatal Nurs. 25 (1) (2019) 37–40.

[3] S.L.W. Fei, K.S. Chew, S. Pawi, L.T. Chong, K.L. Abdullah, L.T. Lim, F. Rintika, Systematic review of the effect of reflective materials around a phototherapy unit on bilirubin reduction among neonates with physiologic jaundice in developing countries, J. Obstet. Gynecol. Neonatal Nurs. 47 (6) (2018) 795–802.

[4] T.M. Slusher, L.T. Day, T. Ogundele, N. Woolfield, J.A. Owa, Filtered sunlight, solar powered phototherapy and other strategies for managing neonatal jaundice in low-resource settings, Early Hum. Dev. 114 (2017) 11–15.
[5] S. Kara, Z. Yalniz-Akkaya, A. Yeniaras, F. Örnek, Y.D. Bilge, Ocular findings on follow-up in children who received phototherapy for neonatal jaundice, J. Chin. Med. Assoc. 80 (11) (2017) 729–732.
[6] T. Colbourn, C. Mwansambo, Sunlight phototherapy for neonatal jaundice—time for its day in the sun? Lancet Glob. Health 6 (10) (2018) e1052–e1053.
[7] T.M. Slusher, L.T. Day, T. Ogundele, N. Woolfield, J.A. Owa, Corrigendum to "Filtered sunlight, solar powered phototherapy and other strategies for managing neonatal jaundice in low-resource settings", Early Hum. Dev. 128 (2019) 121.
[8] K. Sheth, L. Tuyisenge, V.K. Bhutani, Does provider access to technology improve health care? Evidence from a national distribution of phototherapy in Rwanda, Semin. Perinatol. 45 (1) (2021) 151359.
[9] A.A. Raba, A. O'Sullivan, J. Miletin, Prediction of the need for phototherapy during hospital stay in preterm infants by transcutaneous bilirubinometry, Early Hum. Dev. 146 (2020) 105029.
[10] P.W. Chang, W.M. Waite, Evaluation of home phototherapy for neonatal hyperbilirubinemia, J. Pediatr. 220 (2020) 80–85.
[11] S. Kato, O. Iwata, Y. Yamada, H. Kakita, T. Yamada, T. Sugiura, H. Togari, H. Nakashima, Standardization of phototherapy for neonatal hyperbilirubinemia using multiple-wavelength irradiance integration, Pediatr. Neonatol. 61 (1) (2020) 100–105.
[12] M. Basiri, Mouhamad, M. Esmaeili, S. Khosravan, S.J. Mojtabavi, Effects of foot reflexology on neonatal jaundice: a randomized sham-controlled trial, Eur. J. Integr. Med. 38 (2020) 101173.
[13] Y. Qian, Q. Lu, H. Shao, X. Ying, W. Huang, Y. Hua, Timing of umbilical cord clamping and neonatal jaundice in singleton term pregnancy, Early Hum. Dev. 142 (2020) 104948.
[14] G. Korkmaz, F.I. Esenay, Effects of massage therapy on indirect hyperbilirubinemia in newborns who receive phototherapy, J. Obstet. Gynecol. Neonatal Nurs. 49 (1) (2020) 91–100.
[15] M.T.A. Sampurna, R. Etika, M.T. Utomo, S.A.D. Rani, A. Irzaldy, Z.S. Irawan, K.A. Ratnasari, An evaluation of phototherapy device performance in a tertiary health facility, Heliyon 6 (9) (2020), https://doi.org/10.1016/j.heliyon.2020.e04950, e04950.
[16] M.C. Juarez, A.L. Grossberg, Phototherapy in the pediatric population, Dermatol. Clin. 38 (1) (2020) 91–108.
[17] Z. Iliyasu, Z. Farouk, A. Lawal, M.M. Bello, N.S. Nass, N.S. Nass, M.H. Aliyu, Care-seeking behavior for neonatal jaundice in rural northern Nigeria, Public Health Pract. 1 (2020) 100006.
[18] K.-H. Chu, S.-J. Sheu, M.-H. Hsu, J. Liao, L.-Y. Chien, Breastfeeding experiences of Taiwanese mothers of infants with breastfeeding or breast milk jaundice in certified baby-friendly hospitals, Asian Nurs. Res. 13 (2) (2019) 154–160.
[19] E.H. Tham, E.X.L. Loo, A. Goh, O.H. Teoh, F. Yap, K.H. Tan, K.M. Godfrey, H. Van Bever, B.W. Lee, Y.S. Chong, L.P.-C. Shek, Phototherapy for neonatal hyperbilirubinemia and childhood eczema, rhinitis and wheeze, Pediatr. Neonatol. 60 (1) (2019) 28–34.

CHAPTER FOURTEEN

Nanoparticle-based plasmonic devices for bacteria and virus recognition

Suleyman Asir[a], Monireh Bakhshpour[b], Serhat Unal[c], and Adil Denizli[b]
[a]Department of Materials Science and Nanotechnology Engineering, Near East University, Nicosia, Turkish Republic of Northern Cyprus, Turkey
[b]Department of Chemistry, Hacettepe University, Ankara, Turkey
[c]Department of Infectious Disease and Clinical Microbiology, Hacettepe University, Ankara, Turkey

14.1 Introduction

Without biomedical instrumentation and assistive devices for medical diagnostics, it will be challenging to practice in the healthcare field [1, 2]. Biomedical instrumentation and assistive devices for medical diagnostics play a significant role in preserving and enhancing our health and quality of life [1]. Research in this area will continue for the foreseeable future, and there will be ever greater need for these devices, especially as bacterial and viral pathogens cause numerous diseases of inflammation and sometimes even death [3]. Given early successes in the production of vaccinations and drugs over infectious diseases, different and multidrug-resistant infections are increasingly evolving. In addition [4], current bacterial and virus diagnostic methods are inadequate and time consuming, particularly in developing countries, where it remains difficult to find suitable services [5–7]. There is, therefore, a greater need to develop instruments throughout the context of biomedical diagnostics that are fast, portable, user-friendly, and cost-effective tools with quick response times and that are suitable for mass production. Biosensors give the opportunity to achieve those requirements through a multidisciplinary approach of materials science, nanotechnology, chemistry, and medical sciences [3, 8–10].

Biosensors and point-of-care tools are bound to enhance patient outcomes. Biosensor devices can be used as relatively inexpensive disposable point-of-care devices, or they can be used to monitor implantable devices continuously [11]. Early and accurate diagnosis of illness is critical for successful early diagnosis and patient survival. Biomarkers may be tracked

Modern Practical Healthcare Issues in Biomedical Instrumentation
https://doi.org/10.1016/B978-0-323-85413-9.00012-8

noninvasively in specimens, such as saliva and exhaled breath condensate as well as in blood and interstitial fluid via smart stickers, through the instrumentality of biosensors incorporated into wearable devices [12].

The focus of this chapter is on recent developments in the bacterial and virus detection and quantification using nanoparticle- (NP) based plasmonic biosensors. An overview of biosensors, including bioreceptors and transducers, and the nanoparticle-based systems currently used to detect the selected bacteria and viruses will be presented. The final section will address some problems and strategies for the future in this area.

14.2 Biosensors

Biosensors and related technologies are rapidly transforming our potential methods to combat disease like no other time in the field. Clinical measurements of analytes will no longer be conducted exclusively within the laboratory of clinical chemistry. Some measurements, such as blood sugar level measurements, by nonprofessional people in different places have become a common part of our daily life. Major growth continues to be seen in the international market for biomedical biosensors as innovation for engineering materials improves [13, 14]. According to a report from Grand View Research, the latest projections have already shown that the international market for biosensor devices could top USD 36 billion by 2027, largely due to rising demand for point-of-care technologies, or even an increase in chronic infections and general healthcare concern [14].

Biosensors are instruments which can be used to detect contaminants of biological samples such as blood or urine, the concentration of harmful gasses in the soil, or the existence of specific components of any other sample [3, 15–17].

The material of concern examined is named the analyte. As shown in Fig. 14.1, biosensors are composed of three parts: (1) a bioreceptor, which recognizes analyte; (2) a transducer, which converts the process of biorecognition into an accessible output; and (3) a read-out part, which transforms an electrical signal into a readable output [18]. A bioreceptor is required for a biosensor to work, which is the first part where it attaches or responds to the analyte. A meaningful bioreceptor must be very specific and enable bonding only to the molecule of interest. Secondly, a transducer must also be engineered and assembled after the bioreceptor. The transducer is the part that detects recognition of analyte and gives out some kind of signal. Thirdly, it requires a read-out part. This is routinely an electrical device

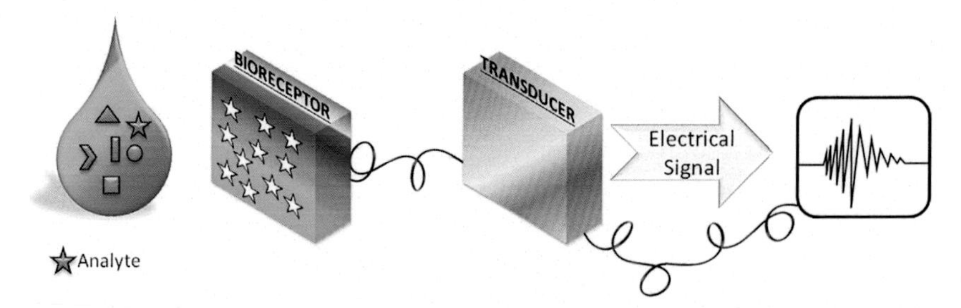

Fig. 14.1 Schematic representation of a biosensor and its components.

that would identify the transducer input signals and demonstrate its strength or intensity in a form that an analyst could comprehend [18, 19].

Biosensors are categorized by the nature of the transducer (like electrochemical biosensors, optical biosensors, piezoelectric biosensors, or thermal biosensors) or the nature of its bioreceptor (like enzyme biosensors; immunobiosensors. Where the bioreceptor is an antibody; DNA biosensors; or whole-cell biosensors). Among these types of transducers, the most common types are electrochemical and optical systems [19–23]. While optical transducers detect an optics-related signal depending on plasmonics, fluorescence, luminescence, infrared, or Raman, electrochemical transducers detection may depend on measurement of impedance, potential, or current.

14.3 Nanoparticle-based plasmonic biosensors

A new area of study recently appeared, named plasmonics, that deals with technology and science relevant to surface plasmon resonance (SPR). After its first use in a real-time determination of a biological sample in the 1990s [24], plasmonic research has now become an important optical biosensing platform for various applications, ranging from pharmaceutical and biological science to diagnostic purposes and global health services, due to its noninvasive character and real-time and label-free detection opportunities.

For centuries, people have identified the distinctive electrical characteristics of plasmonic nanoparticles [25]. Since early civilizations, silver, copper, and gold colloidal nanoparticles have been used by artists to create different colors of stained glass and pottery. In such plasmonic nanoparticles, the beautiful variety of colors result from adjustable optical properties of these nanoparticles. The phenomenon that offers tunable regulation of scattering and absorption of light by nanoparticles can be defined as SPR.

The SPR is an interfacial phenomenon of many particles, in which a precisely defined optical absorption is described by the resonant interaction of moving electrons in thin metal films with incoming photons. This aspect can be used to create optical sensors that can either determine the solution refractive index or detect the presence or formation of the physical adsorption layer in real time [26].

Although many biosensing tools are available, SPR technology makes it less difficult and more precise to observe binding activity at a molecular level than many other kinds of tools [27]. SPR is a simple and straightforward analytical method done by calculating the change in the refractive index on the surface of the metal [28]. A filmy metal layer between two permeable and refractive index media (glass prism and solution) is used in the SPR technique. Commonly, the metal layer is gold or silver. As the plane-polarized light approaches an environment to a greater refractive index over a particular angle, it undergoes full inner reflection. Light is the evanescent wave that goes through the metal film in these circumstances. This wave induces moving electrons on the metal film to form surface plasmons at a particular refraction angle, and a decrease can be observed in the amount of reflected light. This effect is referred to as the SPR and is only detected at a particular angle, named as the resonance angle. The resonance angle can be changed by the attachment of the analyte on the surface of metal film. Changing the surface (by introducing natural or engineered bioreceptors), according to the target analyte forms the fundamental principle of the SPR-based biosensors.

In SPR, the plasmonic substance is usually a gold or silver metal layer that needs a prism to excite the plasmon, whereas nanocolloids or nanofilms are directly excited in another process called localized surface plasmon resonance (LSPR) [28]. As low molecular weight molecules with SPR are often difficult to identify due to the slight difference in the refractive index at binding, the use of nanoparticles is by far the most widespread among many techniques evolved to improve the response of SPR [29].

NP-based devices are much more appealing than conventional micro- and macro-counterparts as they have the advantages of surface-to-volume ratio, direct interaction, intensified physical properties, and reduced size benefits [30–34]. The surface-to-volume increases significantly as the particle size is decreased. Consequently, the linkage of the analyte with the NP-based device is far greater, resulting in a better signal response. The NP-based device's nanoscale dimension is comparable to that of the contact biomolecules, which provide the ability to interact directly with molecular structures. Intensified physical properties, such as increased fluorescence quantum yields or surface-enhanced Raman scattering (SERS), will significantly intensify the signal, resulting in a drastically improved sensitivity. NP size reduction often brings low cost and decreased analysis-time and measurement-volume, which cannot be accomplished with traditional biosensors.

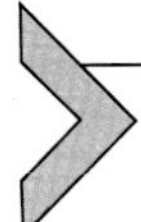

14.4 Applications of nanoparticle-based plasmonic biosensors in bacterial and virus detection

In one study, Ozgur et al. designed a sensitive, selective, and stable SPR system for the detection of *Escherichia coli* (*E. coli*). They used silver nanoparticles for increasing the limit of detection and the limit of the quantification value. This combination demonstrated a sensitive system for the detection of microorganisms [35]. They used an amino acid-based functional monomer for prepared specific cavities on the surface of the SPR ship without using labeling. The characterization studies of this work were done using scanning electron microscopy, atomic force microscopy, water contact angle, and ellipsometer measurements. They used zeta-sizer for reporting the dimensions of silver nanoparticles. The size of nanoparticles was demonstrated 31.5 nm. The selectivity studies of the nanoparticles-combined SPR sensor was reported using *Salmonella* sp. *strains*, and *Staphylococcus* sp. They showed a selective detection for *E. coli* compared to the other microorganisms in the 1.5×10^4 CFU/mL–1.5×10^5 CFU/mL concentration range of the bacteria. Finally, the limit of the detection value was reported at 0.57 CFU/mL.

In another study, Gur et al. used a surface imprinted technology for preparing a selective and sensitive SPR chip for detection of *E. coli*. They designed an Au nanoparticle-based selective SPR sensor for completing a diagnosis of urinary tract infections. The shape and size of microorganisms were caused to obtain a selective binding site on the SPR chip. In the polymerization process, the specific and selective cavities were obtained for the microorganisms with the shape and size memory of the microorganisms. Thus, the selective binding site was created for the selective detection of microorganisms. The rod-shaped and Gram-negative *E. coli* can be detected via this sensor system [16]. They used metal chelator technology with Cu^{2+} metal ions for preparation of the *E. coli*-imprinted SPR chip. The limit of the detection value was reported as 1 CFU/mL in this study. The characterization studies of this work were done using scanning electron microscopy, atomic force microscopy, contact angle, and ellipsometer measurements.

Also, the selectivity of this sensor system was reported against *Pseudomonas aeruginosa*, *Klebsiella pneumonia*, and *Staphylococcus aureus*. They showed that selective cavities were developed on the surface of the SPR sensor; they also successfully reported the reusability of this sensor.

Zhao and others used Pickering emulsion polymerization water-soluble CdTe quantum dots (QDs) to obtain a molecular-imprinted, highly sensitive sensor system for the detection of *Listeria monocytogenes* [36]. Firstly, they treated *Listeria monocytogenes* by acryloyl-functionalized chitosan with QDs to form a bacteria–chitosan network as the water phase. This was then stabilized in an oil-in-water emulsion comprising a cross-linker, monomer, and initiator, causing recognition sites on the surface of microspheres embedded with CdTe QDs. The result of the study showed that the MIP microspheres were enabled to selective capture of the target bacteria via recognition cavities. The visual fluorescence was used to show the sensitive detection of *Listeria monocytogenes* by quenching. They developed a sensitive fluorescence approach for rapid, efficient, and real-time detection of bacteria in the food samples.

In another study, surface imprinting was used for developing receptor-based QDs via molecularly imprinted polymeric-coated materials. Here, signal-transducing materials, CdSe/ZnS, were used. Then these QDs materials were utilized for the detection of a group of autoinducers from Gram-negative bacteria, *N*-acyl-homoserine lactones. *N*-acyl-homoserine lactones offer quorum signaling molecules. This study demonstrated that the imprinting process could be successfully created and that the sensing system was a rapid, selective, and sensitive detection method. The imprinting efficiency was shown using non-imprinting QDs [37].

Another study about these areas was developed by Chen et al. They prepared an electrochemiluminescence sensor for the detection of *E. coli*. They utilized surface-imprinted technology to obtain nitrogen-doped graphene QDs. They showed the methodology for the preparation of polydopamine surface-imprinted polymer-based nanofilm. Therefore, they used electro-polymerized technology for direct electro-polymerization of *E. coli* and dopamine. According to this study, *E. coli* was successfully detected by the electrochemiluminescence signal of QDs. Additionally, they used water samples for the detection of *E. coli* using this novel and sensitive system [38].

Wang et al. used SPR nanosensors for highly rapid, selective, and sensitive recognition of pathogenic bacteria. They prepared a system that depends on the spectroscopic measurement of grating-coupled long-range surface plasmons. They used a magnetic nanoparticle combination in this study. Fig. 14.2 shows the schematic of the chip, optical setup, and used

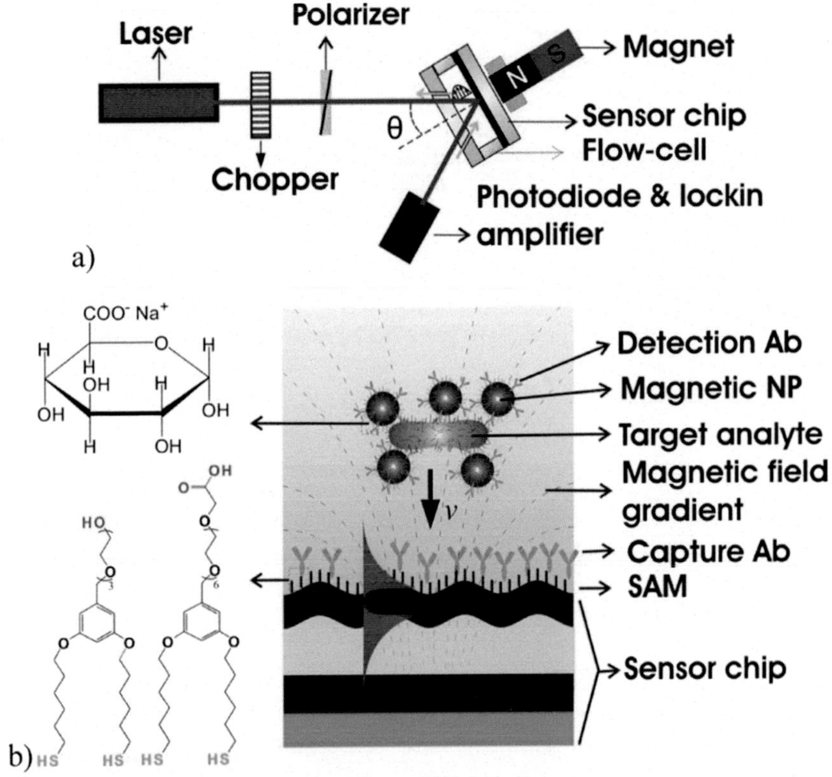

Fig. 14.2 (A) Optical setup using the GC-LRSPs. (B) Utilized surface architecture for bacterial pathogens detection via MNP enhanced assay. Magnetic nanoparticle (MNP), grating-coupled long-range surface plasmons (GC-LRSPs). *(From Y. Wang, W. Knoll, J. Dostalek, Bacterial pathogen surface plasmon resonance biosensor advanced by long range surface plasmons and magnetic nanoparticle assays. Anal. Chem. 84 (19) (2012) 8345–8350).*

surface structure for the bacterial detection. They reported a 50 CFU/mL detection value for *E. coli* O157:H7 [39].

Paul et al. improved with a bioconjugated Au nanoparticles SERS probe for the sensitive detection of West Nile virus (WNV) and Dengue virus (DENV). They showed a highly sensitive Raman fingerprint for detection of both viruses via the antiflavivirus 4G2 antibody, which is conjugated with Au nanoparticles. They reported 10 plaque-forming units (PFU)/mL of viruses via Au nanoparticles-based SERS. Selectivity of our probe was validated using another mosquito-borne chikungunya virus as a negative control. The data showed that the SERS enhancement factor was recorded as 10^4 times selective [40]. Fig. 14.3 shows the schematic preparation of the construction of antiflaviviral Au nanoparticles and detection of viruses.

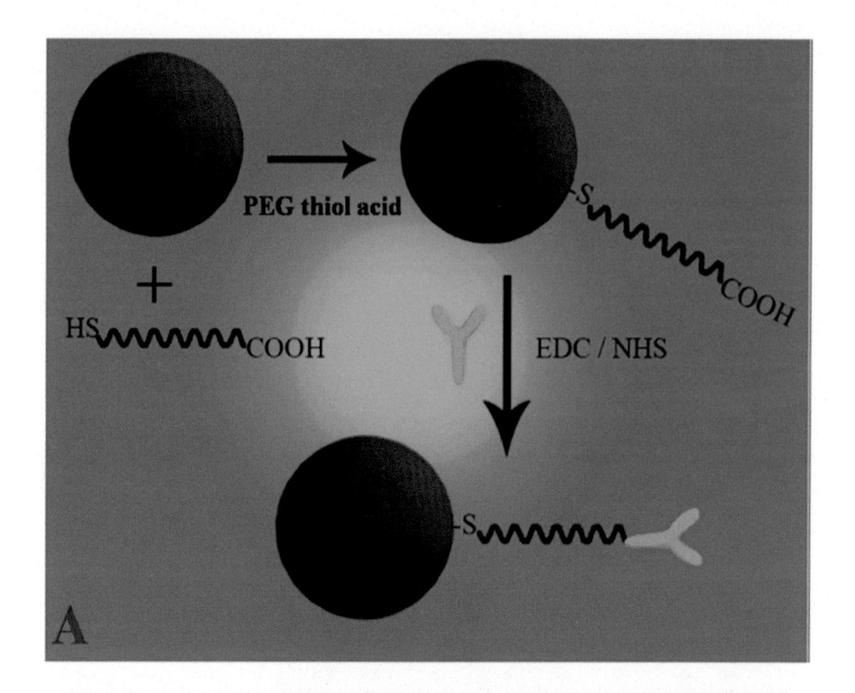

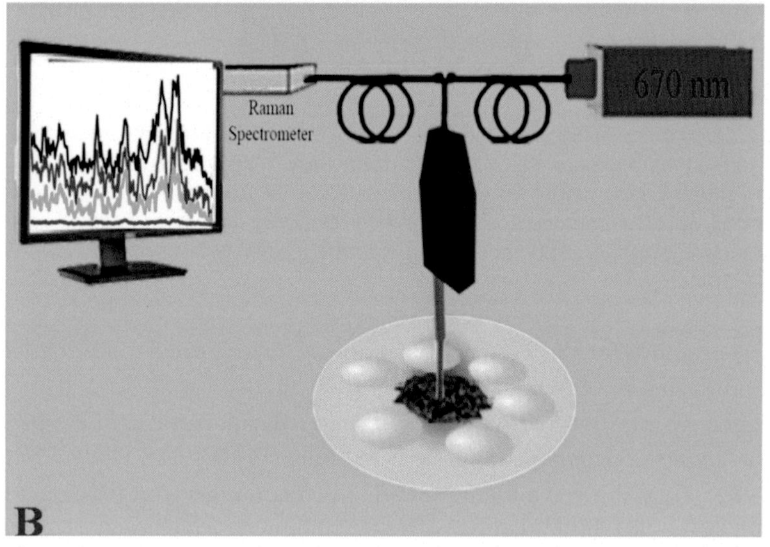

Fig. 14.3 (A) Preparation of the construction of antiflaviviral- (4G2) coated gold nanoparticles. (B) Detection of viruses via Raman fingerprinting [40].

Zhu et al. prepared an LSPR effect-based sensor via the hybrid Au and Ag nanoparticles array for detection of *Staphylococcus aureus* enterotoxin B [41]. The direct enterotoxin B detection via the LSPR sensor was obtained at the low concentration level (ng/mL) while other SPR sensor systems need amplification steps for sensitive detection of enterotoxin B. The summary of nanoparticles-based sensor technology used for the detection of microorganisms is shown in Table 14.1.

To give an example from literature, Eom et al. used functional Au nanoparticles for highly sensitive and selective detection of an oseltamivir-resistant (pH1N1/H275Y mutant) virus. They designed functional gold nanoparticles via modification of the surfaces of nanoparticles with malachite green isothiocyanate and oseltamivir hexylthiol as a receptor. They showed sensitive detection of pH1N1/H275Y mutant viruses via SERS and the naked eye (Fig. 14.4) [45].

Reinhard et al., used R, G, and B pixel intensities to quantify color coordinates in the HSV, CIE L*a*b*, and rgb chromaticity color spaces. They synthesized Au nanoparticles in sizes ranging between 18 and 115 nm. Then, the prepared Au nanoparticles were attached to the functionalized glass

Table 14.1 The summary of nanoparticles-based sensor systems for the detection of microorganisms.

Nanoparticles-based sensing	Function of nanoparticles	Microorganisms	Detection value	Ref.
PAMAM-NH_2 dendrimer	Signal amplification	*Escherichia coli*	10^4 CFU/mL	[42]
Dye dope silica nanoparticles	Biosensing system	*Escherichia coli O157:H7*	1 CFU/mL	[43]
Au nanoparticles	Biosensing system	*Staphylococcus aureus enterotoxin B*	0.1 ng/mL	[41]
Ag nanoparticles	Biosensing system	*Salmonella Enteritidis* insertion element gene	1 ng/mL	[44]
Au nanoparticles	Signal amplification	*Escherichia coli*	58.2 ± 1.37 pg/mL	[45]
Polymeric nanoparticles	Biosensing system	*Escherichia coli*	0.57 CFU/mL	[35]
Au nanoparticles	Biosensing system	*Escherichia coli*	1 CFU/mL	[16]
Polymeric nanoparticles	Biosensing system	*Enterococcus faecalis*	2×10^4–1×10^8 CFU/mL	[10]

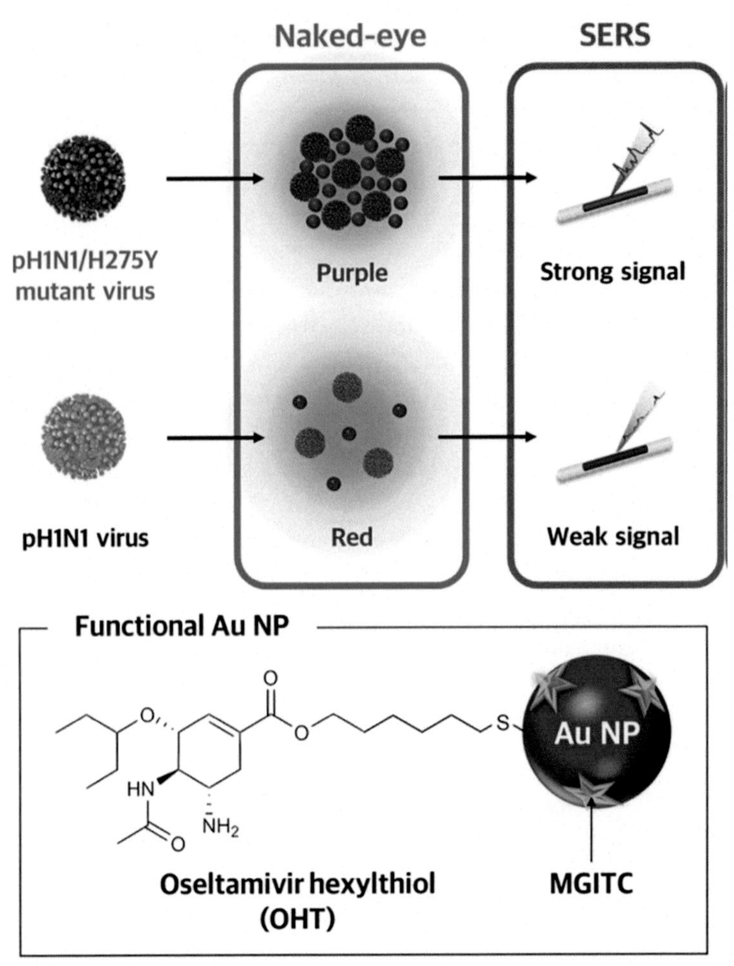

Fig. 14.4 Schematic preparation of naked-eye and SERS dual-mode sensing procedures for virus with functional Au-NPs [45].

coverslips. The sensor was characterized by SEM analyses. They demonstrated that hue (H) is the most sensitive color parameter, with a change per refractive index unit of 0.71 and a figure of merit of 183 RIU^{-1} for sensors comprising 115 nm diameter nanoparticles. They showed that the size of Au nanoparticles relying on hue measurement plays an important role to achieve a linear sensor response. They reported general design rules for preparing colorimetric sensors and demonstrated how to use their sensors with a smartphone [46]. The production of Au nanoparticle sensor chips was shown in Fig. 14.5.

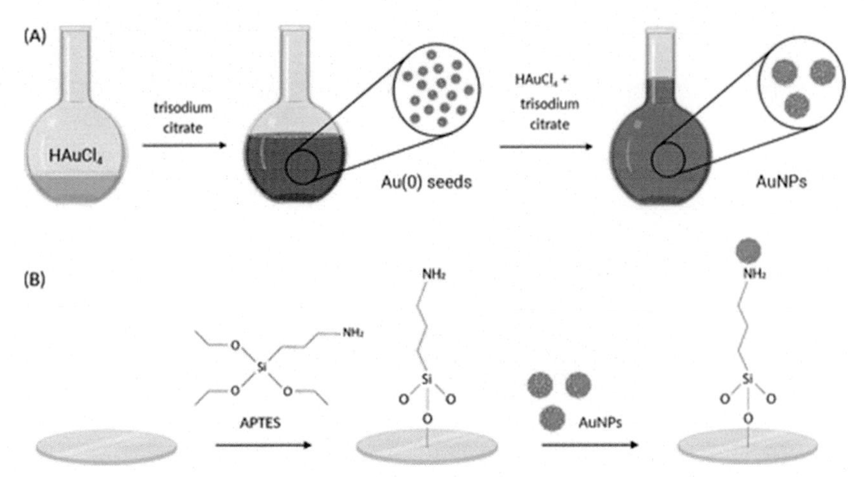

Fig. 14.5 Production of Au nanoparticle sensor chips [46].

Gokturk et al. synthesized acridine orange-loaded poly(2-hydroxyethyl methacrylate) polymeric nanoparticles by mini-emulsion polymerization for cell labeling. Also, the toxicity of the nanoparticles was investigated by MTT [3-(4,5-dimethylthiazol-2-yl)-2,5-diphenyltetrazolium bromide] assay. The effect of acridine orange-loaded polymeric nanoparticles on cell viability was appraised by MTT assay, based on the reduction of the number of metabolically active cells. It has been observed that fluorescent-labeled nanoparticles localized mainly in the nucleus of L929 cells and did not affect the cell viability for 72 h [47].

Inci et al. showed a dual-sensing modality of the nanoplasmonic sensor by integrating antibody-based gold (Au) nanoparticles on a surface for label-free detection. They used sandwiching biotargets between nanoparticles on the top and the bottom. They used the nanoplasmonic sensor for detection of the Hepatitis B virus (HBV) in serum samples that had been isolated from infected individuals, spanning a clinical relevant range between 10 and 10^5 IU/mL. Figs. 14.6 and 14.7 were shown the schematic of workflow in HBV detection assay and detection of HBV on the sensing surface, respectively [48].

Pathogenic microorganisms threaten the health of livestock, wildlife, and human populations. The detection of these pathogens is a major step for the successful treatment and diagnostics of infectious animal diseases [49]. Biosensors as early detection are cost-effective and ultrasensitive tools, as cited in the literature, for the detection of microorganisms. A large number of optical sensors, including the SPR and LSPR, have been widely used in the

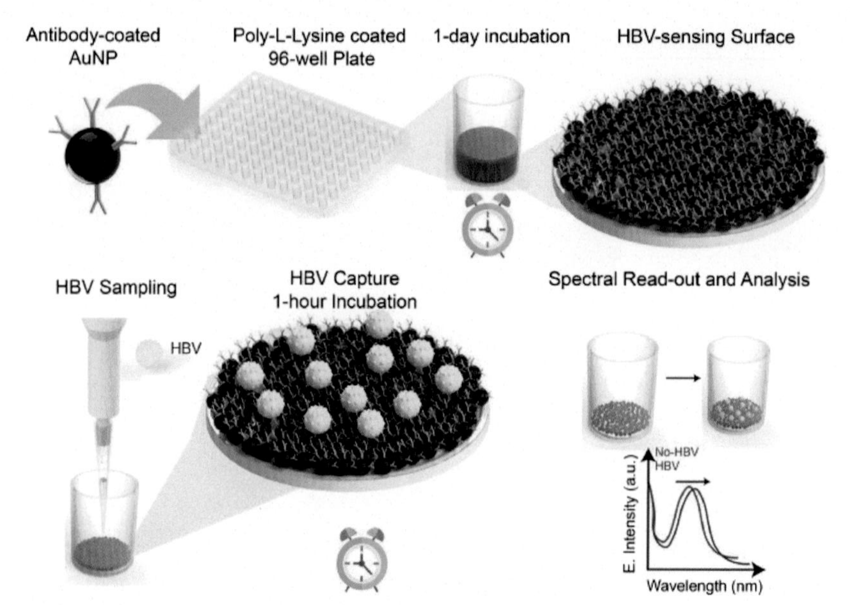

Fig. 14.6 Schematic of workflow in HBV detection assay [48].

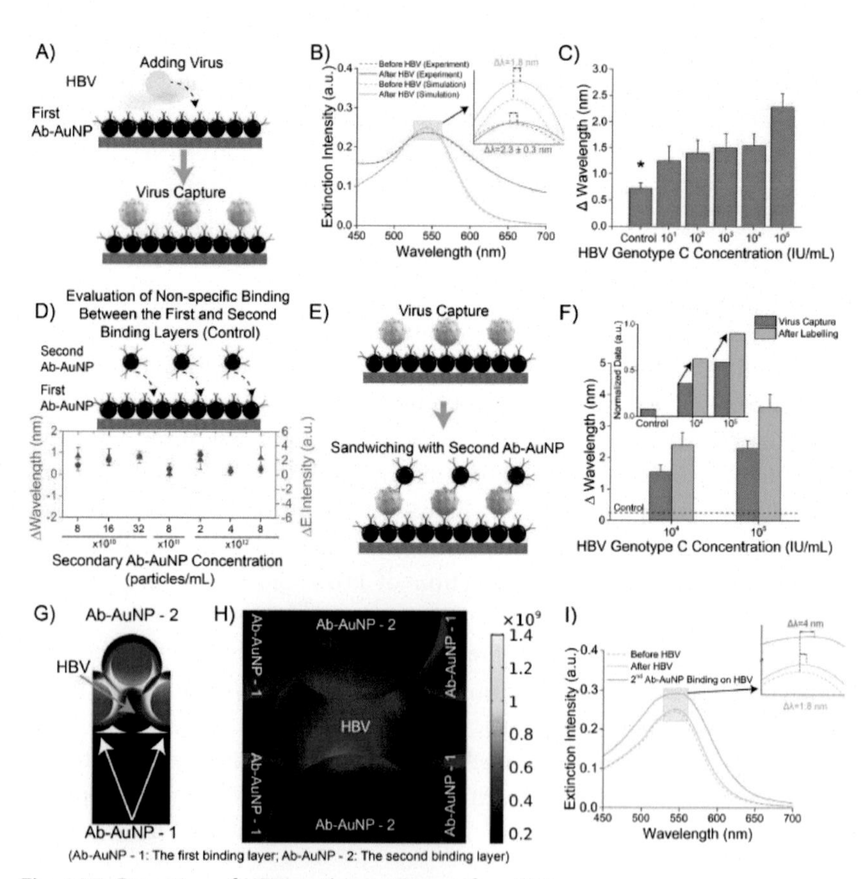

Fig. 14.7 Detection of HBV on the sensing surface [48].

detection of viral strains, such as those associated with SARS, H1N1, influenza, and MERS in the laboratory [50]. A large number of recent studies are focused on developing the detection methods of coronavirus via biotechnology-based methods, and these studies report about the structure of coronavirus as well as the detection of coronavirus with a PCR system. Because of this urgency related to this COVID-19 pandemic, the diagnostics of coronavirus require cheap, early, and reliable detection methods [51].

Xia et al. prepared a strategy for simple and proper staining to target components of the protective surface layers of the endospores. They used thioflavin T, an amyloid staining dye in their study. They reportedly incubated bacillus endospore suspensions with thioflavin T for a short time then showed fluorescent images with unique contrast ratios while increasing the fluorescence of thioflavin T and the accumulation of thioflavin T in the endospore [52, 53].

Ramírez-Navarro et al. prepared a simple, efficient, and cheap colorimetric test for the detection of the NS1 protein of dengue virus in infected serums via magnetic (Fe_3O_4) nanoplatforms. They coupled these nanoparticles with anti-NS1 antibodies. They added the magnetic nanoparticles into infected serum then showed that the nanoparticles were conjugated with the protein; therefore, the nanoparticles were easily separated using an external magnetic field. They showed the results in the ELISA system. According to their results, a color was changed to blue in the presence of a protein that was induced by reaction with the Perls reagent. This result fits with the system of an antibody–antigen conjugate that confirms infection [54].

14.5 Conclusions

Many plasmonic devices based on nanoparticles have been developed in recent years for the identification of bacteria and viruses, while plasmonic characteristics of these devices have been enhanced and improved to maintain efficient outputs. Combination of nanoparticles and plasmonic-based measurement systems can be used widely as high-precision biosensing devices for modern practical healthcare issues. Such developments have also shown a large range of uses, primarily for medicinal and biomedical measurements, but also for environmental, industrial, nutritional, and agricultural applications.

The research on plasmonic-based nanoparticles is oriented toward industrial and commercial applications from academic laboratories, where the majority of the work discussed in this chapter is brought. We may think of clinical practices as a big commercial opportunity, but to reach the clinical industry for nanoparticle-based plasmonic devices, further research is required. This is promising that a significant range of uses have been documented, including clinical sample analysis utilizing nanoparticle-based plasmonic instruments, which has proven the broad applicability of such tools to a variety of essential clinical problems. Additionally, validating a new clinical procedure needs specific analytical criteria to be tested. Measuring a clinical test's variation is calculated by the responsiveness inside and within a sample, the routine result, and the response that differs across various locations and expertise. In order to accredit these kinds of plasmonic devices for the identification of bacteria and viruses for clinical purposes, significant numbers of patients would need to be tested, and greater correlation research would be needed.

This review explicitly demonstrates that a huge number of plasmonic devices are compact and are getting simpler and smaller, allowing real sample analysis, and examples of plasmonic sensors based on nanoparticles were also effective in biomedical applications. Hence, bacterial and virus recognition systems focused on nanoparticle-based plasmonic devices can be commonly used as extremely sensitive instruments for modern practical healthcare issues.

References

[1] E.P. Balogh, B.T. Miller, J.R. Ball, Committee on Diagnostic Error in Health Care, Board on Health Care Services, Institute of Medicine, Improving Diagnosis in Health Care, National Academies Press (US), Washington, DC, 2014.
[2] A. Heidari, A novel approach to future horizon of top seven biomedical research topics to watch in 2017: Alzheimer's, Ebola, hypersomnia, human immunodeficiency virus (HIV), tuberculosis (TB), microbiome/antibiotic resistance and endovascular stroke, J. Bioeng. Biomed. Sci. 7 (1) (2017) e127.
[3] Y. Saylan, Ö. Erdem, S. Ünal, A. Denizli, An alternative medical diagnosis method: biosensors for virus *detection*, Biosensors 9 (2) (2019) 65.
[4] Y. Saylan, A. Denizli, Molecularly imprinted polymer-based microfluidic systems for point-of-care applications, Micromachines 10 (766) (2019) 1–14.
[5] S.M. Fletcher, M.-L. McLaws, J.T. Ellis, Prevalence of gastrointestinal pathogens in developed and developing countries: systematic review and meta-analysis, J. Public Health Res. 2 (1) (2013) 42.
[6] A. Denizli, in: A. Denizli (Ed.), Affinity Sensors, 50th Year of Hacettepe University Book Series, Hacettepe University Press, Ankara, 2018.
[7] T.D. Gootz, The global problem of antibiotic resistance, Crit. Rev. Immunol. 30 (1) (2010).
[8] Y. Saylan, A. Denizli, Virus detection using nanosensors, in: B. Han, V.K. Tomer, N.T. Anh, A. Farmani, P.K. Singh (Eds.), Nanosensors for Smart Cities, Elsevier, Amsterdam, 2020, pp. 501–511 (Chapter 30).

[9] N. İdil, M. Hedström, A. Denizli, B. Mattiasson, Whole cell based microcontact imprinted capacitive biosensor for the detection of Escherichia coli, Biosens. Bioelectron. 87 (2017) 807–814.
[10] Ö. Erdem, Y. Saylan, N. Cihangir, A. Denizli, Molecularly imprinted nanoparticles based plasmonic sensors for real time Enterococcus faecalis detection, Biosens. Bioelectron. 126 (2019) 608–614.
[11] S. Akgönüllü, H. Yavuz, A. Denizli, SPR nanosensor based on molecularly imprinted polymer film with gold nanoparticles for sensitive detection of aflatoxin B1, Talanta 219 (1) (2020) 121219.
[12] N. Idil, M. Bakhshpour, I. Perçin, A. Denizli, Molecularly imprinted nanosensors for microbial contaminants, in: M. Inamuddin, A. Asiri (Eds.), Nanosensor Technologies for Environmental Monitoring, Nanotechnology in the Life Sciences, Springer, Cham, 2020.
[13] P. Maresova, L. Hajek, O. Krejcar, M. Storek, K. Kuca, New regulations on medical devices in Europe: are they an opportunity for growth, Adm. Sci. 10 (1) (2020) 16.
[14] Grand View Research, Acrylonitrile Market Size, Share & Trends Analysis Report by Application (Acrylic Fibers, Adiponitrile, Styrene Acrylonitrile, Acrylonitrile Butadiene Styrene, Acrylamide, Carbon Fiber), by Region, and Segment Forecasts, 2020–2027, 2020. Market Research Report.
[15] M. Çalışır, M. Bakhshpour, H. Yavuz, A. Denizli, HbA1c detection via high-sensitive boronate based surface plasmon resonance sensor, Sens. Actuators B Chem. 306 (2020) 127561.
[16] I. Perçin, N. Idil, M. Bakhshpour, E. Yılmaz, B. Mattiasson, A. Denizli, Microcontact imprinted plasmonic nanosensors: powerful tools in the detection of Salmonella paratyphi, Sensors 17 (2017) 1375.
[17] S.D. Gür, M. Bakhshpour, A. Denizli, Selective detection of Escherichia coli caused UTIs with surface imprinted plasmonic nanoscale sensor, Mater. Sci. Eng. C 104 (2019) 109869.
[18] M. Bakhshpour, A. Denizli, Highly sensitive detection of Cd (II) ions using ion-imprinted surface plasmon resonance sensors, Microchem. J. 159 (2020) 105572.
[19] Y. Saylan, Ö. Erdem, F. İnci, A. Denizli, Advances in biomimetic systems for molecular recognition and biosensing, Biomimetics 5 (2) (2020) 20.
[20] D. Çimen, N. Bereli, S. Günaydın, A. Denizli, Detection of cardiac troponin-1 by optic biosensors with immobilized anti-cardiac troponin-I monoclonal antibody, Talanta 219 (2020) 121259.
[21] D.M. Alhaj-Qasem, M.A.I. Al-Hatamleh, A.A. Irekeola, M.F. Khalid, R. Mohamud, A. Ismail, F.H. Mustafa, Laboratory diagnosis of paratyphoid fever: opportunity of surface plasmon resonance, Diagnostics 10 (2020) 438.
[22] M. Bakhshpour, A.K. Piskin, H. Yavuz, A. Denizli, Quartz crystal microbalance biosensor for label-free MDA MB 231 cancer cell detection via notch-4 receptor, Talanta 204 (2019) 840–845.
[23] Y. Saylan, F. Yılmaz, E. Özgür, A. Derazshamshir, H. Yavuz, A. Denizli, Imprinting of macromolecules for sensors applications, Sensors 17 (4) (2017) 898.
[24] R. Karlsson, A. Michaelsson, L. Mattsson, Kinetic analysis of monoclonal antibody-antigen interactions with a new biosensor based analytical system, J. Immunol. Methods 145 (1–2) (1991) 229–240.
[25] W.P. Halperin, Quantum size effects in metal particles, Rev. Mod. Phys. 58 (3) (1986) 533.
[26] T. Vo, A. Paul, A. Kumar, D.W. Boykin, W.D. Wilson, Biosensor-surface plasmon resonance: a strategy to help establish a new generation RNA-specific small molecules, Methods 167 (2019) 15–27.

[27] A.N. Naimushin, S.D. Soelberg, D.U. Bartholomew, J.L. Elkind, C.E. Furlong, A portable surface plasmon resonance (SPR) sensor system with temperature regulation, Sens. Actuators B Chem. 96 (1–2) (2003) 253–260.
[28] S.E. Kim, M.V. Tieu, S.Y. Hwang, M.H. Lee, Magnetic particles: their applications from sample preparations to biosensing platforms, Micromachines 11 (2020) 302.
[29] A. Shalabney, I. Abdulhalim, Sensitivity-enhancement methods for surface plasmon sensors, Laser Photonics Rev. 5 (4) (2011) 571–606.
[30] J. Li, N. Wu, in: J. Li, N. Wu (Eds.), Biosensors Based on Nanomaterials and Nanodevices, CRC Press, 2013.
[31] I. Willner, B. Willner, Biomolecule-based nanomaterials and nanostructures, Nano Lett. 10 (10) (2010) 3805–3814.
[32] W. Putzbach, N.J. Ronkainen, Immobilization techniques in the fabrication of nanomaterial-based electrochemical biosensors: a review, Sensors (Switzerland) 13 (4) (2013) 4811–4840.
[33] N. Sanvicens, C. Pastells, N. Pascual, M.P. Marco, Nanoparticle-based biosensors for detection of pathogenic bacteria, TrAC: Trends Analyt. Chem. 28 (11) (2009) 1243–1252.
[34] J.M. Pingarrón, P. Yáñez-Sedeño, A. González-Cortés, Gold nanoparticle-based electrochemical biosensors, Electrochim. Acta 53 (19) (2008) 5848–5866.
[35] E. Özgür, A.A. Topçu, E. Yılmaz, A. Denizli, Surface plasmon resonance based biomimetic sensor for urinary tract infections, Talanta 212 (2020) 120778.
[36] X. Zhao, Y. Cui, J. Wang, J. Wang, Preparation of fluorescent molecularly imprinted polymers via pickering emulsion interfaces and the application for visual sensing analysis of Listeria monocytogenes, Polymers (Basel) 11 (6) (2019) 984.
[37] J.D. Habimana, J. Ji, F. Pi, E. Karangwa, J. Sun, W. Guo, F. Cui, J. Shao, C. Ntakirutimana, X.J. Sun, A class-specific artificial receptor-based on molecularly imprinted polymer-coated quantum dot centers for the detection of signaling molecules, N-acyl-homoserine lactones present in gram-negative bacteria, Anal. Chim. Acta 15 (1031) (2018) 134–144.
[38] S. Chen, X. Chen, L. Zhang, J. Gao, Q. Ma, Electrochemiluminescence detection of Escherichia coli O157:H7 based on a novel polydopamine surface imprinted polymer biosensor, ACS Appl. Mater. Interfaces 9 (6) (2017) 5430–5436.
[39] Y. Wang, W. Knoll, J. Dostalek, Bacterial pathogen surface plasmon resonance biosensor advanced by long range surface plasmons and magnetic nanoparticle assays, Anal. Chem. 84 (19) (2012) 8345–8350.
[40] A.M. Paul, Z. Fan, S.S. Sinha, Y. Shi, L. Le, F. Bai, P.C. Ray, Bioconjugated gold nanoparticle based SERS probe for ultrasensitive identification of mosquito-borne viruses using Raman fingerprinting, J. Phys. Chem. C 119 (41) (2015) 23669–23675.
[41] S. Zhu, C.L. Du, Y. Fu, Localized surface plasmon resonance-based hybrid Au-Ag nanoparticles for detection of Staphylococcus aureus enterotoxin B, Opt. Mater. (Amst) 31 (2009) 1608.
[42] A.N. Naimushin, S.D. Soelberg, D.K. Nguyen, L. Dunlap, D. Bartholomew, J. Elkind, J. Melendez, C.E. Furlong, Detection of Staphylococcus aureus enterotoxin B at femtomolar levels with a miniature integrated two-channel surface plasmon resonance (SPR) sensor, Biosens. Bioelectron. 17 (2002) 573.
[43] J. Ji, J.A. Schanzle, M.B. Tabacco, Real-time detection of bacterial contamination in dynamic aqueous environments using optical sensors, Anal. Chem. 76 (2004) 1411.
[44] D. Zhang, D.J. Carr, E.C. Alocilja, Fluorescent bio-barcode DNA assay for the detection of Salmonella enterica serovar Enteritidis, Biosens. Bioelectron. 24 (2009) 1377.
[45] G. Eom, A. Hwang, D.K. Lee, K. Guk, J. Moon, J. Jeong, J. Jung, B. Kim, E.-K. Lim, T.K. Kang, Superb specific, ultrasensitive, and rapid identification of the oseltamivir-

resistant H1N1 virus: naked-eye and SERS dual-mode assay using functional gold nanoparticles, ACS Appl. Bio Mater. 2 (3) (2019) 1233–1240.
[46] I. Reinhard, K. Miller, G. Diepenheim, K. Cantrell, W.P. Hall, Nanoparticle design rules for colorimetric plasmonic sensors, ACS. Appl. Nano Mater. 3 (5) (2020) 4342–4350.
[47] I. Göktürk, V. Karakoç, M.A. Onur, A. Denizli, Characterization and cellular interaction of fluorescent-labeled PHEMA nanoparticles, Artif. Cells Nanomed. Biotechnol. 41 (2) (2013) 78–84.
[48] F. Inci, M.G. Karaaslan, A. Mataji-Kojouri, P.A. Shah, Y. Saylan, Y. Zeng, A. Avadhani, R. Sinclair, D.T.Y. Lau, U. Demirci, Enhancing the nanoplasmonic signal by a nanoparticle sandwiching strategy to detect viruses, Appl. Mater. Today 20 (2020) 100709.
[49] J. Vidic, M. Manzano, N. Jaffrezic-Renault, Advanced biosensors for detection of pathogens related to livestock and poultry, Vet. Res. 48 (2017) 11.
[50] N. Bhalla, Y. Pan, Z. Yang, A.F. Payam, Opportunities and challenges for biosensors and nanoscale analytical tools for pandemics: COVID-19, ACS Nano Article ASAP 14 (7) (2020) 7783–7807.
[51] R. Jalandra, A.K. Yadav, D. Verma, N. Dalal, M. Sharma, R. Singh, A. Kumar, P.R. Solanki, Strategies and perspectives to develop SARS-CoV-2 detection methods and diagnostics, Biomed. Pharmacother. 129 (2020) 110446.
[52] B. Xia, S. Upadhyayul, V. Nuñez, P. Landsman, S. Lam, H. Malik, S. Gupta, M. Sarshar, J. Hu, B. Anvari, G. Jones, V.I. Vullev, Amyloid histology stain for rapid bacterial endospore imaging, J. Clin. Microbiol. 49 (8) (2011) 2966–2975.
[53] Z. Qin, R. Peng, I.K. Baravik, X. Liu, Fighting COVID-19: integrated micro- and nanosystems for viral infection diagnostics, Matter 3 (3) (2020) 628–651.
[54] R. Ramírez-Navarro, P. Polesnak, J. Reyes-Leyva, U. Haque, J.C. Vazquez-Chagoyán, M.R. Pedroza-Montero, M.A. Méndez-Rojas, A. Angulo-Molina, A magnetic immunoconjugate nanoplatform for easy colorimetric detection of the NS1 protein of dengue virus in infected serum, Nanoscale Adv. 2 (2020) 3017–3026.

Index

Note: Page numbers followed by *f* indicate figures, and *t* indicate tables.